IBERIAN AND LATIN AMERICAN STUDIES

The Spanish National Research Council and the Shaping of Science in Franco's Spain

Other titles in the series

Spain's Eighteenth-Century Periodical Press: Antonio Calladares de Sotomayor and the *Semanario Erudito*
Madeline Sutherland-Meier

Against Borges: Four Women Writers of his Inner Circles
Patricia N. Klingenberg

The Catalan Crisis: Between Spanish Liberal Democracy and State (dis)Unity
Sergi Auladell Fauchs

The Other Catalans: Representations of Immigration in Catalan Literature
Edited by Josep-Anton Fernàndez

Postcolonial Spain: Coloniality, Violence and Independence
Edited by Helena Miguélez-Carballeira

The Space of Latin American Women Modernists
Camilla Sutherland

Paradise in Hell: Alcohol and Drugs in the Spanish Civil War
Jorge Marco

Motherhood and Childhood in Silvina Ocampo's Works
Fernanda Zullo-Ruiz

Jesus of Nazareth in the Literature of Unamuno
C. A. Longhurst

Latin America and Existentialism: A Pan-American Literary History (1864–1938)
Edwin Murillo

Spain is different? Historical memory and the 'Two Spains' in turn-of-the-millennium Spanish apocalyptic fictions
Dale Knickerbocker

Blood, Land and Power: The Rise and Fall of the Spanish Nobility and Lineages in the Early Modern Period
Manuel Perez-Garcia

IBERIAN AND LATIN AMERICAN STUDIES

The Spanish National Research Council and the Shaping of Science in Franco's Spain

FERNANDO GARCÍA NAHARRO

UNIVERSITY OF WALES PRESS
2026

www.uwp.co.uk

British Library CIP
A catalogue record for this book is available from the British Library.

ISBN 978-1-83772-399-7
e-ISBN 978-1-83772-400-0

Published by arrangement with The University of Texas at Austin.

This project was supported by the European Union's NextGenerationEU programme, the Spanish Ministry of Science, Innovation and Universities and the State Research Agency, RYC2021-033006-I

MIX
Paper | Supporting responsible forestry
FSC® C013604

Typeset by Geethik Technologies
Printed by CPI Antony Rowe, Melksham, United Kingdom

To my brother

Contents

List of Figures

List of Tables

Note on Sources and Acknowledgements

This work is the revised, updated English-language version of the book *Palabras, actores, objetos y lugares. El CSIC y la demarcación de la ciencia en la España franquista (1939–1966)*, published in 2022 by Editorial CSIC, the publishing house of the Spanish National Research Council. This edition contains a new prologue and epilogue, although the document appendices from the original edition are not included, and the critical apparatus has been reduced to the strict minimum to meet the criteria for the new edition. For in-text citations throughout the book, the Harvard-style referencing system is used, with additional endnotes containing the citations for archival and official reports for each chapter.

The contents of this book benefited greatly from an intellectual exchange with renowned international experts in the History of Science and Scientific Communication, particularly Jonathan R. Topham and Graeme Gooday (University of Leeds), Mario Biagioli (†) (University of California, Davis) and Agustí Nieto-Galan and Xavier Roqué (Universitat Autònoma de Barcelona). To these scholars I send my deepest thanks for providing me with the opportunity to take part in research residencies and work alongside them. Many other academics and researchers also contributed by providing information and documents that unquestionably improved the contents of the manuscript. The book was also enhanced by the recommendations and suggestions made by various peer reviewers. I send my gratitude and recognition to them all, although their names are too numerous to properly list here. Special mention, however, is due to my esteemed colleagues who took the time to read the early drafts of the book, either completely or partially: Jesús A. Martínez Martín, Carmen Romero Bachiller, Jean-François Botrel, Leoncio López-Ocón Cabrera, Elena Hernández Sandoica, Luis Enrique Otero Carvajal, Albert Presas i Puig and Fernando J. García Selgas. I would also like to take this opportunity to acknowledge Richard Cleminson, Professor of Hispanic Studies at the University of Leeds, who was so kind as to write the prologue to this edition, and Pamela J. Lalonde, the translator of this work, for helping me to find my voice in the English language. From here, to everyone, I once again send my most sincere thanks.

Series Editors' Foreword

Over recent decades the traditional 'languages and literatures' model in Spanish departments in universities in the United Kingdom has been superseded by a contextual, interdisciplinary and 'area studies' approach to the study of the culture, history, society and politics of the Hispanic and Lusophone worlds – categories that extend far beyond the confines of the Iberian Peninsula, not only in Latin America but also to Spanish-speaking and Lusophone Africa.

In response to these dynamic trends in research priorities and curriculum development, this series is designed to present both disciplinary and interdisciplinary research within the general field of Iberian and Latin American Studies, particularly studies that explore all aspects of Cultural Production (inter alia literature, film, music, dance, sport) in Spanish, Portuguese, Basque, Catalan, Galician and indigenous languages of Latin America. The series also aims to publish research in the History and Politics of the Hispanic and Lusophone worlds, at the level of both the region and the nation-state, as well as on Cultural Studies that explore the shifting terrains of gender, sexual, racial and postcolonial identities in those same regions.

Prologue

The history of scientific endeavour as an accretional undertaking, removed from and devoid of the circumstances of its own generation, has long been cast aside as a suitable epistemology to understand how science is conceived and performed *in situ*. This book, *The Spanish National Research Council and the Shaping of Science in Franco's Spain* by Fernando García Naharro, fulfils the promise of a new departure for studies of science that develop not only a network approach that takes into account the ways in which knowledge is disseminated and received, but also examines the relationship between people, ideas, material objects and the very institutions and laboratories where this practice takes place.

In common with the work of Bruno Latour and Steve Woolgar's *Laboratory Life: The Social Construction of Scientific Facts* (1979), García Naharro focuses on the minutiae as well as the 'big picture' of scientific thought and practice and in doing so the everyday artefact, the reading room, the scientific journal and the habitus of each scientific workplace fall under the author's scrutiny. While the former work was revolutionary in its dissection of the sometimes authoritarian and often teleological dynamics of the laboratory, this volume on the Spanish National Research Council (CSIC) stands as a volume that rescues what the author calls the 'insignificant', and situates it within a historical genealogy of scientific policy under the first three decades of the dictatorial regime of Francisco Franco (1939–75). As such, as the reader may suspect, the 'insignificant' turns out to be replete with profound ramifications for the development of ideas, techniques and the careers of many people.

In examining such dynamics, the work reveals itself as important for a much deeper understanding of the constraints – but also the opportunities for those 'on side' or merely having adopted a neutral stance – presented by the regime for the development of science. Indeed, one of the many merits of this book is to show that despite the restrictions and cruel and sometimes arbitrary purges undertaken from the early days of the tribunal of 'Political Responsibilities' and the general repressive atmosphere, which, it will be recalled, effectively exiled Spanish scientists of the standing of Gregorio Marañón for some years, avenues were explored to undertake scientific research that resulted in important national and international advances.

There are three axes around which such developments took place and that strike the reader as particularly important. While certain areas of research and practice, such as mental health and psychiatry were practised in suffocating surroundings, rooted in Christian moral frameworks and, at least in the years immediately after the Civil War, bound to constitutionalist notions (as individuals such as Carlos Castillo del Pino relates in his eloquent *Pretérito imperfecto* (1997)), the new regime, like the Portuguese dictatorship to a lesser degree, if it sought anything, wished for the name of Spain to be writ large on the international scientific stage. As such, it funded overseas research grants and even though the home-bred atmosphere in which the applications for these were conceived may have been stultifying, once researchers began to work abroad, eyes and ears were opened to new ideas and Spanish researchers could test out work of their own making against international standards and freer conversations.

Paradoxically though it may seem, many of these researchers were women. Women did research, published and were in receipt of these grants. The loosening of the ideological constraints imposed by the regime in respect of scientific endeavour, as with other parts of the social matrix including the presence of women in traditional male-dominated workplaces and gender roles, began to take place in the late 1950s as the restrictive 'taken-for-granted' rules of scientific labour, journal writing and the very questions that were to be asked of science were transformed as a new humanistic Catholicism began slowly to take root in Spanish soil. New institutes and new academic reviews were created as part of the regime's hubris, but also the genuine scientific verve of those who were permitted roles in the development of the dictatorship's scientific architecture began to demand to be heard.

This is not a story, however, of a regime that was benign in any sense of the word. Overseeing these three pillars of scientific undertaking – internationalisation, the presence of women in science, and the proliferation of scientific journals – was the Spanish National Research Council. Founded as a nationalist body on the ashes of the republican, progressive Junta para Ampliación de Estudios, the CSIC was authoritarian, top-down and technocratic. As such, it endeavoured as a process of scientific anthropophagy to absorb the '*Arbor Scientiae* that sprouted in Spain after the Civil War during the time of alleged financial and industrial self-sufficiency' (p. 5). Within this new framework, some prospered while many were forced to leave Spain or suffer immeasurable acts of repression, personally and in their work.

In addition to examining with the aid of a vast range of sources from annual reports to the discourse within a variety of scientific and technical journals, the mechanics whereby the state and its offspring, the CSIC, managed to achieve this, pressing questions are posed. What ethical position, if any, is it possible to adopt under such circumstances? How can science be performed under such a regime? Perhaps even more importantly, what are the possibilities for resistance and new ideas despite the overarching control, repression and marginalisation of many practitioners? Such questions are mobilised by García Naharro to shed light on a vital ingredient of the Franco dictatorship, which both undergirded it but also provided the seeds for its questioning in the 1960s. This study of scientific practice under fascist and dictatorial regimes has shown how, in accordance with separating out authorised and non-permitted science and practitioners, the regime consolidated an isolated nationalist system faithful to a renewed expression of Neo-Thomism in the early years after the Civil War. This gave way, as change began to operate in society, to a modern and more international form of science, often protagonised to some degree by women, which was increasingly inserted into a market-driven consumer-based model that would draw on discourses that would open up shafts of light in both science and society that the regime would increasingly find difficult to close down.

Richard Cleminson
University of Leeds
Spring 2025

Introduction

A Plea for the Study of the Insignificant

> I swear by God and by Spain that the contents of the six packages being sent from the 'Alonso Barba' Institute, two for Switzerland, two for Germany and two for Italy, contain work booklets published in *Anales de la Real Sociedad Española de Física y Química*.[1]

Thus read a short note dated August 1943 and written, presumably, by the doctor of Chemical Sciences, Julio Guzmán Carrancio (1883–1956), director of the Electrochemistry Section and secretary of the National Institute of Physics and Chemistry (*Instituto Nacional de Física y Química*, INFQ, Instituto Rockefeller) at the Board for Advanced Studies and Scientific Research (*Junta par la Ampliación de Estudios e Investigaciones Científicas*, JAE) during the 1930s. This renowned scientist was not only an interim professor of electrochemistry at the Faculty of Sciences at the Universidad Central from 1929 to 1930, but was also a grant recipient at the Wilhelm-Ostwald-Institut in Leipzig, Germany from 1912 to 1913 – where he conducted research with the chemist Carl Drucker (1876–1959) in the company of Enrique Moles (1883–1953) – and a fellow at the Bureau of Standards in Washington DC and the University of California, Berkeley in 1919. Nonetheless, in the summer of 1936, without warning, he was among the employees expelled from the INFQ for being deemed hostile to the Red cause. By the 1940s, however, he was one of the few so-called survivors of the former Rockefeller Institute (INFQ) and had a position working at the newly created Alonso Barba Institute of the Spanish National Research Council (*Consejo Superior de Investigaciones Científicas*, CSIC)

(Fernández Terán, 2014: 337–44; Otero Carvajal and López Sánchez, 2012: 338–9).

The Spanish Civil War (1936–9), of course, drastically disrupted nearly every aspect of life in Spain. Nonetheless, after the coup d'état, Julio Guzmán decided not to leave the country and, when the war was over, he began to collaborate once again with his colleague Julio Palacios (1891–1970) on reorganising the damaged National Institute of Physics and Chemistry (which, after the formal constitution of the CSIC, ended up being absorbed into the new Alfonso el Sabio Board). The former vice president of the JAE, José Casares Gil (1886–1961), became the director of the two new institutes within the board, Alonso Barba and Alonso de Santa Cruz, which now divided research into physics and chemistry, although both were still located in the old Rockefeller Building.[2] Such was the situation in that summer of 1943, when, from the building of the former National Institute of Physics and Chemistry, Guzmán posted – by written affidavit and with his contact information on the back – the printed records of a number of scientific works published by the communication channel of the Spanish Royal Society of Physics and Chemistry. The packages were posted to three recipients: the diplomat Guillermo Nadal y Blanes of the Spanish legation in Bern, Switzerland; diplomat and cultural adviser Ignacio Oyarzabal in the Spanish embassy in Berlin, Germany; and a 'Mr Harrison', secretary in the Spanish embassy in Rome, Italy.[3]

Earlier that same year, and again through diplomatic representatives, a number of embassies had made significant donations of books and journals to the CSIC,[4] rendering the six packets believed to have been sent by Guzmán quite small in comparison, an almost insignificant gesture. However, that action epitomises many of the subjects and themes examined throughout this book. That is also exactly why these introductory pages avail themselves of both this apparently inconsequential act and the tentative, fragmentary and marginal data that comprise it, bringing to the forefront, and thus concisely explaining, the symbolic importance of gestures like this one, not only in the thematic composition of this book, but also in the systematic way in which this work approaches its subject.

In other words, this is perhaps a slightly unconventional study, but that aims to supplement an abundant historiography, albeit one that is still incomplete and subject to debate: the history of the Spanish National Research Council, the sciences in general and the production of scientific practices and knowledge during the dictatorship of Francisco Franco. The book engages with an extensive bibliography, both at the beginning and throughout each of the four parts into which its contents are organised, in an attempt to give greater coherence and consistency to each section. In addition, and in the interest of providing more specific information regarding the primary subject, the book consciously does not include the social, biomedical or health sciences. This makes it possible to delve more deeply into the sphere of the applied sciences, and in doing so mirror the interests of the dictatorship and of that time. This involves using a perspective different to the one commonly adopted in the specialist literature on the topic, applying instead an interdisciplinary approach that builds bridges between history and the sociology of science, and the history of publishing, books and reading. This theoretical/methodological pairing provides the necessary interpretive keys to successfully reach the aim of the book: to analyse, with some degree of theoretical rigour, the criteria used in Franco's Spain to constitute both science/s and the definition of scientific authority, while simultaneously determining the social actors would could – or could not – be involved in the production and validation of the scientific and theoretical knowledge of the era, in both practical and discursive terms.

Ultimately, then, the aim of this book is to use what is learned from each of the topics examined in great detail in each section, to reconstruct – and grapple with – the criteria at play when the time came to demarcate (Gieryn, 1983; Lamont and Molnár, 2002; Laudan, 1983; Woolgar, 1991;) and define what science was and was not, later protecting it from unwanted participants and interferences (Shapin, 1992). Moreover, this book primarily approaches the subject from the perspective of the printed products of science, which are often lost to oblivion. In this particular case, the scientific and technical publications of the CSIC are viewed in the context of their production and circulation, examining both the categories associated with their field and the procedures behind their selection and the construction of the status of scientific knowledge in the spatiotemporal framework of Franco's Spain. These publications, then, are understood here not only as textual channels or media, but as objects, as cultural products whose formal dimension – design, format and/or language – become an essential part of attracting – or deterring – readers, as well as shaping the ways in which the texts are read, if indeed they are. The reading public, whether consumers or merely spectators, are nonetheless always the final judge of publications that are textually and physically marked by their formal configuration and the places where they can be acquired or consulted. In this case, the surrounding context was a dictatorial regime that labelled these publications as acceptable or objectionable according to their content, even if in practice these descriptors took on their own meaning in the hands of the readers into which the works fell (Ruiz Bautista, 2004 and 2008).

Accordingly, by returning agency to both written and printed products, as well as to their producers, their readers and the contexts in which they circulated, this book makes an explicit plea on the part of what has generally been considered a secondary question but which, following the approach proposed herein, can be quite revealing. Equally enlightening is the fact that, during the early 1940s Julio Guzmán was one of the few authors in *Anales de la Real Sociedad Española de Física y Química* who had published during the period before the Civil War and continued to do so (Palao, Valera, López Fernández and Marset, 1984: 44, n. 18). Moreover, while Guzmán spoke at the conference of the Central Association of Industrial Engineers in the 1930s, in May 1940, he continued to lecture on electrochemistry before the School of Industrial Engineers. This random data might suggest that little or nothing had really changed for him, just as on that note dated August 1943, Guzmán appeared as the secretary of the National Institute of Physics and Chemistry by written affidavit on the letterhead stationery, with his name, former telephone number and address of the Rockefeller Building at 119 Serrano Street, Madrid.[5]

But, as is well known, the reality was quite different. By that time, the JAE had ceased to exist and Guzmán would never again hold the position of secretary of an institute that, moreover, was no longer the same. This was seen clearly in the address on those six packages, that of the Alonso Barba Institute, which was included with the emblem of the CSIC,[6] the *Arbor Scientiae* that sprouted in Spain after the Civil War during the time of alleged financial and industrial self-sufficiency. These years of repression and the purging of so-called secular, liberal anti-Spain, of scarcity, rationing and the black market (Aróstegui, 2012; De Riquer, 2010; Gil Pecharromán, 2008; Viñas, 2012) clearly permeate the pages of this book, although it tackles a longer period, a complex and changing period – albeit with some continuities – marked by the denouement of the Second World War and the resulting new economic, scientific and social order (Camprubí, Roqué and Saez de Adana, 2018; Krige, 2006; Santesmases Navarro de Palencia, 2009).

Words (or the Disperse Fragments that Nurture the Homogenous Appearance of the Official Narrative about Science)

Michel de Certeau once said that 'making history' always leads to writing, to the meaningful practice that transforms material from the past into a produced text (Certeau, 2006: 19–20). In the case at hand, the result is a sort of fragmented staging of the past that nonetheless finds its unity – in other words, the relationship between the content and the form of the book itself – in the very way that the subject is approached. This is largely due to the fact that this investigation is based on the tensions inherent in 'saying' and 'making', without denying the potential encroachment of discourses into practices. In this case, the discourses come from distinct figures in specific institutions that should not be uncoupled from the production process or political praxis that accompany them.

In this respect, the object of study in this section comprises the speeches delivered by the first president of the Spanish National

Research Council and Spanish Minister of National Education, José Ibáñez Martín (1896–1969) – the 'pure politician at the service of the ideals of the regime' – who always counted on the close collaboration of José María Albareda (1902–66) – the 'great scientist in his specialty of edaphology' (Muñoz Ruiz, 1990: 19) – as secretary-general and inspiration for the work of the CSIC (Malet, 2008: 214–35; Sánchez Ron, 1999: 329–71). It was, in fact, Alborada himself who, in April 1940, wondered whether scientific research could 'continue interrupted, its publications suspended, its exchanges without resuming, while the red emigrees begin to publish and those who have stayed here think that we are not capable of organizing anything without those who left'.[7] He continued:

> The reality is that national research can be organized with a range and depth highly superior to all that of the past. Thirty years broadening our studies under the JAE has left us with fundamental branches where no work was attempted; not a thought given to Philosophy, Geography, Technology. But it is not for us now to criticize many of their undertakings.[8]

In 1942, Ibáñez Martín asserted that 'it is not only a radical change in the orientation of the guiding principles of knowledge; the very execution of research must be rectified'.[9] From this perspective, Julio Guzmán's change of status at the CSIC, in part, reflected all these issues: he no longer even served as the director of a section. In the meantime, the secretaries of the new physics and chemistry institutes were, respectively, the naval engineer José María Otero Navascués (1907–83) – 'a devout Catholic and loyal member of the rebel army' (Herran and Roqué, 2012: 88; Pérez Fernández-Turégano, 2012) – and the chair professor of chemistry and member of Catholic Action, Manuel Lora-Tamayo (1904–2002) (Lora-Tamayo, 1993), who was also a member of the National Education Council and would become board member and president of the Spanish Royal Society of Physics and Chemistry, thanks to Ibáñez Martín. In other words, pre-war affronts, resentments and ideological affinities also played their role in the process of dismantling the foundations of the so-called Silver Age of Spanish science. The well-known purge of scientists and academics (Barona Vilar, 2010; Canales Serrano and Gómez-Rodríguez 2017; Claret Miranda, 2006; López Sánchez, 2013; Otero Carvajal, 2006) shaped a new scientific reality in Spain, established under the renewed directives of a violent, authoritarian regime.

For this very reason, and in line with some of the most recent studies undertaken to comprehend the genesis and development of institutional science in that complex and dictatorial Spain (Camprubí, 2017; Delgado Gómez-Escalonilla and López García, 2019; Janué i Miret and Presas i Puig, 2021; Nieto-Galan, 2019; Otero Carvajal, 2017; Sánchez Ron, 2021; Santesmases Navarro de Palencia, 2017), the first part of this book is dedicated to understanding and delineating the self-declared official concept of science as put forward at that time from the leading Spanish scientific institution: the Spanish National Research Council. What were they talking about when they discussed science and technology? To what and to whom were they referring? How did they do it and why? These and other questions are addressed in Part I: Words, in an in-depth analysis of the speeches given by José Ibáñez Martín, CSIC president from 1939 to 1967 and minister of national education from 1939 to 1951, which are used to examine the processes behind the construction of the official image of science handed down from the highest levels of the Franco regime. These speeches, loaded with praise for the regime and rhetorical exaltation, also contained lines of argument that were common to the fascisms of their time (Eco, 1995; Gómez and Canales, 2009). Moreover, they were delivered in communicative contexts that were conducive to their common pragmatic purpose: to outline an ideal concept of science, understood as purely national, purportedly unitary and harmonious, Catholic in tradition and supposedly free, albeit subordinate to 'the living realties facing the state'.[10]

Although these ideas and arguments related to scientific research underwent some interesting changes over time, they were nonetheless set down in texts – a coherent fabric of written and oral pronouncements laden with citations, descriptions and half-hidden references – that, far from restricting their influence to the merely rhetorical sphere, positioned and marked differences in practice as well. In short, at play in Ibáñez Martín's speeches was the imposition of an avowedly legitimate – legitimised, authorised and official – view of not only science, but also its professionals, products and spaces. This is, therefore, an official image of science that, as this book explores, constructed and directly spoke to the subject / expert as legitimised by the regime, always placing them in a differential position on the scales of power and responsibility.

Actors (or the External Shapers of their Names and the Collective Image of the Scientist)

However, to limit this exercise to the official discourse and reduce its subjects and contexts to mere textual fictions would mean avoiding the murmur, the constant, low rumble of silenced things (Latour, 2007: 83–133) that motivated the writing of this book. To a large extent, this silence also accompanied the historical figure of Julio Guzmán, although his name appeared regularly in *Anales de la Real Sociedad Española de Física y Química* during the 1930s. By contrast, in the 1940s he published only two articles in that journal; one on the electroanalysis of Spanish minerals in 1941, and a second on the electroanalysis of nickel and its application to Spanish garnierite in 1942. The articles were co-authored by his colleagues 'P. Pardo' and Luis Sardaña, respectively, scientists who remained on the staff of the Alonso de Santa Cruz Institute for a short time afterwards. It is not known whether these were the works that Guzmán presumably posted during the summer of 1943, but it is known that the works done in his institute that ended up published in *Anales* during these years were monopolised by the names of Manuel Lora-Tamayo and Antonio Rius Miró (1890–1973), the secretary and director, respectively, of the CSIC's Alonso Barba Institute.[11] Like Guzmán and Lora-Tamayo, Antonio Rius Miró was also a former staff member of the JAE, who became the editor of *Anales* and a key figure in Chemical Engineering in Franco's Spain, always operating under the protection of his former student, secretary-general of the CSIC, José María Albareda (Nieto-Galan, 2019: 150–1).

Why, when most of the associates of the JAE were purged, did a chosen few become notable members of the CSIC? And in this context, what premises underlay the figure of the legitimate scientist? In the search for new answers that, one might say, loosen the lacings of the 'disciplinary matrix' (Kuhn, 2006: 312–20), and in line with recent works that aim to rise above the dichotomy between science and ideology (Camprubí, 2017; Florensa, 2017; Florensa and Roqué, 2021; Presas i Puig, 2021), Part II: Actors, studies the socio-discursive conception of the field of science in Franco's Spain. Once again, this analysis owes a great deal to a semiotic perspective – but without being bound to a strictly linguistic approach – that makes it possible to determine how the purportedly neutral and aseptic space of science became a place of privilege for the male/subject/knower, the *vir modestus* (Haraway, 2004), along with the few women who also figured out how to find their place and even make a name for themselves.

The pages that comprise Part II are to a large extent dedicated to the study of the so-called 'scientist', understood as an actor and author. This entails studying the characteristics associated with the figure of the scientist as legitimised by the regime, as well as everything concerning the process of constructing these figures, approached here through their reflection in the pages of scientific publications: printed products validated by scientific institutions that circulated through social spaces inhabited, in turn, by people and institutions recognised within the field of science. Of course, the authority of the publications and their authors at the time came from an established scientific order to which these very publications then also contributed, only accepting works that deserved to be called 'scientific' in the interest of the dominant criterion (Bourdieu, 1975: 29–31). Science is also made from and with publications, and the writers of scientific works, by this measure, are established as scientific authors. For that reason, because of the role that they play in what is written, the names of the authors – men and women of science (although the women did not necessarily refer to themselves using female pronouns) – always have a meaning beyond the mere name (Daston and Sibum, 2003). Who writes and from where, how the 'author function' (Biagioli, 2006; Foucault, 1998; Wilson, 2004) is assigned and materialised, are all examined through some figures that are often silenced: the women of science. The female authors of the scientific and technical articles published in the CSIC's main periodical, the *Revista de Ciencia Aplicada,* which was tasked with technical and industrial research missions – subject to the fundamental guidelines of the unity of science and serving the national interest – from within the Juan de la Cierva Board, whose establishment, period of splendour and later decline would ensue during the years of the dictatorship covered in this book (López García, 2008: 95–103).

This investigation also brings to light the practical influence of the official discourse of the regime about women in science in a Spain where, by decree, men and women were educated and socialised separately. What class of woman, then, had the legitimacy and independent status needed to be able to choose to be scientific authors under the Franco regime? Following the path broken by the studies already done on the relationship between science and gender in Spain, this part of the book introduces the women of science with their full names, their brilliant backgrounds and careers and equipped, moreover, with a specific social capital spread across networks of recommendations and social – and conjugal – relations with the legitimated – male – scientists of their time (Pimentel Igea, 2020: 291–342; Romero de Pablos, 2016 and 2017; Romero de Pablos and Santesmases Navarro de Palencia, 2017; Santesmases Navarro de Palencia, 2000). In doing so, and with the names of the female researchers at the CSIC who filled research positions and attended exchanges abroad and training programmes in this authoritarian, dictatorial and sexist Spain, it becomes possible to reconstruct a good part of the 'pertinent features' (Bourdieu, 1992: 230–1) and socio-professional profiles of these women, who published and inhabited the 'peripheral fields' of science, where gaining access would be the chief difficulty for women (Rossiter, 1997), such as engineering and the applied sciences; fields on which the CSIC also concentrated some of its most vigorous efforts (Lora-Tamayo, 1963: 14).

Printed Objects (or the Socio-Material Reality of Scientific Journals that Incorporate Meanings and Help to Shape the Scientific Field)

Writing in the 1960s, Thomas Kuhn observed that 'scientific knowledge, like language, is intrinsically the common property of a group or else nothing at all' (Kuhn, 1962) – hence the importance of studying not only the scientific knowledge produced, but also the groups that create, administer and then incorporate it into the social heritage (Nieto-Galan, 2011). To examine these questions in relation to the scientific community in Franco's Spain, this book engages in a dialogue with authorised studies in this area, mostly approaching these issues from the perspective of the methodological postulates of abstract empiricism (González Blasco, 1980; González Blasco and Jiménez Blanco, 1979). While statistical assertions are always borne in mind, this work steers clear of them, methodologically speaking, looking instead at the 'ways of doing' and the 'processes' that produce inequality (Lahire, 2006: 31–40). This applies to particularly decisive points related to the construction of scientific authority, the configuration – both practical and discursive – of the spaces, agents and objects of science and the constitution of science as, supposedly, autonomous (Biagioli, 1999; Galison and Stump, 1996; Golinsky, 2005; Shapin, 2008 and 2010). In this context, this is all studied based on the written culture and printed production of science, because the real aspiration of this book is to connect written production with the social function and circulation of what is printed, its media and its forms. These theoretical and methodological presuppositions call for a relational analysis that considers the human interactions and motives involved in texts (McKenzie, 2005: 32), in order to understand – in all its complexity – how the printed products of science acquire value and credit in a specific sociocultural context and time (Biagioli and Galison, 2003; Csiszar, 2018; Frasca-Espada and Jardine, 2000; Johns, 1998; Secord, 2000).

The scientific publications that Julio Guzmán apparently took it upon himself to pack up and send abroad in the summer of 1943 were highly valued in Spain as well. All these works were published in the renowned journal put out by the Spanish Society of Physics and Chemistry (*Sociedad Española de Física y Química,* SEFQ) which, after the Civil War, recovered the adjective 'Royal' bequeathed by King Alfonso XIII in his day (Valera Candel and López Fernández, 2001: 195, n. 451). However, this was not the only change in this Spanish scientific society in post-war Spain. During the summer of 1939, its members were purged and a course was set that culminated in the institutional linkage of the SEFQ with the CSIC. From then on, articles written by Spanish scientists such as Enrique Moles, Blas Cabrera, Francisco Giral or Antonio Madinaveitia y Tabuyo (1890–1974) would not again appear in the SEFQ journal's pages. These illustrious men – and names – well known by Guzmán were now reviled. On the other hand, different authors who had also been published in the journal during the Second Republic, such as Julio Palacios, José María Torroja, Julio Garrido and Piedad de la Cierva, continued to work and publish under the aegis of the CSIC's *Arbor Scientiae,* which now extended its branches only for those who, it was said, 'had the right to ask for its shade and protection'.[12]

Shade, protection and also a means of expression; even during the harshest times of the post-war era – with its inherent supply problems – the CSIC researchers were always able to rely on their own books and journals, for which paper was never lacking. The effort was also reflected in successive increases in the budget allocation made by the Franco government to project the image of a robust science, underpinned by the regime's own powerful editorial work and supplied with institutional coverage, dissemination and financing (López Sánchez and Fernández Gallego, 2021). This was an integral part of the unwavering 'social contract for science' (Guston, 2006; Jasanoff, 2004) established between the scientific community and the state, in which the latter financed research while supporting the official channels that existed to publish the results (García Naharro, 2015b). Moreover, like the other official centres and institutions under Franco, the CSIC's technical and scientific publications were exempt from the pre-emptive censorship to which the bulk of the editorial production in Spain was subject until 1966 (Martínez Martín, 2015). The fact is that in Franco's Spain, not all publications were equal. As one example, an order issued on 18 February 1941, signed by José Ibáñez Martín, stipulated that it was mandatory for all institutes of secondary and higher education to purchase CSIC publications 'on a preferential basis above the others'; this applied as well as to 'University, Provincial and Lending Libraries'.[13]

The flagrantly unequal conditions and occasionally illegal – and, more often than not, abusive – competition imposed by various official entities (the ministries, CSIC and *Editora Nacional*) on the private initiative of publishers created a kind of structural subordination in the commercial sector in the face of the institutional privileges (García Naharro, 2019). This was established not only through trade inequities, but also by forging links between certain editorial products and publication services in the name of enshrining science, in close liaison with the political sector. All these conditions and constraints determine the ending point for this book: the passage of the Press and Printing Law in 1966 and the consequent lifting of pre-emptive censorship, which closed a chapter that had been open since the Spanish Civil War. During those decades, state regulations stipulated that the official scientific institutions were responsible for approving all works published by private publishing companies via a report issued by the head of the corresponding institution – CSIC institutes and centres, state universities, technical schools – thus, making them, in practice, the ultimate guardians of the scientific and technical works that appeared in published form in the Spanish commercial sector.

The scientific productions of the Spanish National Research Council, then, enjoyed these and other privileges. The CSIC publications are examined in Part III: Printed Objects – although only through a detailed study of one particular editorial product and format: the council's scientific journals. These official technical and scientific publications functioned not only as the institutional locus of scientific authority in Spain, but also as a sort of offering to the Caudillo and printed guarantee of the scientific production of the Franco regime (Canales, 2019; Malet, 2008). Moreover, their function as objects of exchange with other journals from international research centres was of vital importance, in part because these transactions alleviated the financial burden on the libraries associated with the CSIC's institutes. The focus in this section, then, falls on official journals, written largely in Spanish by CSIC researchers and volunteers who were compensated, in many cases, using the funds raised from subscriptions and advertisements on the journal pages. This vast number of journals met both the need to provide the CSIC researchers with a publication outlet and a clear promotional objective: to project a powerful image of Spanish science equipped with productive and influential research facilities. This coveted influence was, however, not always reflected by reality, and the commitment to quantity often simply resulted in the production of duplicates.

Other studies have looked at the rhetoric and language of these types of publications to reveal the intentions of their producers (Halliday and Martin, 1993; Latour 1992: 21–62). The analysis here, on the other hand, seeks to articulate a narrative around the editorial projects, the media for the printed material and the circulation spaces of these journals, elements that also played a determining role when it came to selecting their readers. Part III first breaks the journals down in order to then analyse their constitutive elements, from sections and illustrations to advertisements and the names of many of the different stakeholders involved in the process of shaping them. These textual and material elements, combined with the information provided by their empirical readers, make it possible to reconstruct a portion of the presumed circulation routes of the printed products. In a politically mediated process, these spaces, journeys, uses and forms give meaning to these journals, in addition to their identity and status (Appadurai, 2003: 3–63).

Places (or the Physical and Normative Limits of the Library Space, Understood as a Means of Organisation and Social Interaction)

During the years of the Franco era, not only did some CSIC technical and scientific publications sporadically come out of the cloister of the seminars and libraries to be displayed in the public space as a way, it was said, 'to show people how work is being done in all the spheres of culture',[14] they were also exhibited abroad and exchanged with other scholarly publications. Moreover, they were used to demonstrate to the international scientific community that researchers like Julio Guzmán were still working. Just as the American astronomer Henry Norris Russell (1877–1957) had written from Princeton University, for scientific reasons, during the Spanish Civil War to request offprints of an article by Miguel Catalán published in *Anales de la Sociedad Española de Física y Química* (Sánchez Ron, 1994: 304–5), in the summer of 1943, it seems to be Guzmán who sent a variety of scientific works published in that same journal abroad from the Alonso Barba Institute. Through scientifically motivated acts such as these, however, the Franco dictatorship also shored up its official political rhetoric about a vigorous Spanish scientific community openly communicating with the outside world. Once again, small gestures – combined with larger ones like inviting professors to visit, sending Spanish researchers to attend international conferences and participate in research stays abroad – became active, operational parts of the slow process of gaining international recognition for Spain and its scientific community (Delgado Gómez-Escalonilla, 2007; Sánchez Ron, 2002: 17–31; Santesmases Navarro de Palencia, 2008; Sesma Ladrin, 2021).

As argued throughout this book, it was not only through agreements, programmes and meetings that the precarious transnational scientific and technological relationships were established; the circulation of knowledge and printed products played a role as well. One example of these privileged vehicles of scientific knowledge (Villacorta Baños, 2016) comprised the journals on nuclear energy that the American naval attaché gifted to the

Naval Research Laboratory and Workshop (*Laboratorio y Taller de Investigación del Estado Mayor de la Armada*, LYTIEMA) in 1946 under the leadership of José María Otero Navascués, awakening an interest in nuclear research years before the founding of the Spanish Atomic Research Board (*Junta de Investigaciones Atómicas*, JIA) (Rubio-Varas and De la Torre, 2018: 90). The regime also took advantage of scientific knowledge for propaganda purposes, as in the case of the American donation of a small library to the Board of Nuclear Energy (*Junta de Energía Nuclear*, JEN) (Romero de Pablos, 2018: 63–84).

There is no question that one particularly fascinating facet of printed products is that they are able to fulfil a variety of diverse and unexpected functions, including for their authors and/or owners. For example, both the personal correspondence of Albert Einstein and the nearly 2,400 works in his library – largely books, but also journals – today help to better understand this figure and the scientific canon of his time (Gelfand, 2019; Holton, 2008: 8–11). These works include books by Ludwig Boltzmann (1844–1906), a driving force behind scientific and anti-metaphysical philosophy, as well as Wilhelm Wien, Heinrich Hertz and Max Planck, who also appear – to return to the CSIC – among the hundreds of works on physics and chemistry that belonged to the personal private libraries of Enrique Moles and Blas Cabrera, but were confiscated to become part of the collections in the libraries of the Alonso Barba and Alonso de Santa Cruz Institutes.

The titles of these works, which belonged at the time to two *personae non gratae* in the eyes of the regime, can now be used to reconstruct part of the history of two illustrious men of Spanish science who were maligned under Franco. Likewise, and somewhat paradoxically, these printed products provide information about the scientific readers at the CSIC, the institution that had cast both men aside. These books, both bearers of authority and the printed reflection of the scientific identity (Nye, 1993) of the disciplines practiced by Moles and Cabrera, demonstrate the continuity of the authority of their experts and the circulation of knowledge (Santesmases Navarro de Palencia, 2012: 301). Later they become reflections of the desire to learn among the readers at the CSIC – including Julio Guzmán himself – who consulted the confiscated books. By the same token, the unceasing petitions to the library to purchase foreign publications also challenge the theoretical complete isolation of Spanish science, which never existed in practice.

For all that, and despite the budget shortages that required some requests for reading material to be prioritised over others (and where, once again, the authority of the petitioner played an essential role), the requests that have survived create a small window into the reading experiences of the past for the historian. However, even with this information about the interests in specific works, little is really known about their end use, whether the lists of authors and books, of journals and research articles were employed in the end and under what conditions (García Naharro, 2020a). What is known, on the other hand – in part thanks to the letters sent to various CSIC researchers by individual readers – is that in the grey Spain of the post-war era, beyond the circles of so-called 'high culture', there was a community of readers interested in scientific and technological content. As seen Part IV: Places, in their pursuit of knowledge, these readers used informal means to try to overcome the physical, regulatory and legal obstacles of the regulated spaces of the CSIC libraries. Amid this whole system of prohibitions established by the council in accordance with the book policies deployed in Spain during the Franco period (Andrés, 2012; Larraz 2014; Martínez Rus, 2012 and 2014; Ruiz Bautista, 2005 and 2008), it is possible to glimpse some of what the research community did on a daily basis, something so common and routine that it is usually silenced in the bibliography. However, as this brief introduction has attempted to show – and as the rest of the book will expand on – what are allegedly the most obvious assumptions and insignificant details can provide the basis for considerations in which the apparently most trivial matters become fundamental. Let us begin.

Part I

Words

Chapter 1

José Ibáñez Martín: The Enduring Imprint of the Founder of the Spanish National Research Council

> In the tranquil semi-darkness of the laboratory and the quiet enclosure of the library, the days scarcely make their silent passing heard, and the years are no more than links in an immense and varied endeavour. The Truth exceeds the bounded limits of time, though it is reached by way of temporal paths, and the rhythm of scientific work does not follow the outside rotations of the sun, but lies, rather, in the deep and essential imperatives of the investigative undertaking.[1]

With these words, José Ibáñez Martín (1896–1969), minister of national education and president of the Spanish National Research Council (CSIC), began his speech of 17 April 1950 during the solemn closing session of the commemorative ceremonies to mark the tenth anniversary of the founding of the CSIC. Even then, these concise words demarcated places, recorded times and established the significance of a specific activity – scientific research – in the context of an event to celebrate the intrinsic value of science and its institutions. On that special occasion, in addition to the customary acts that characterised the CSIC's annual plenary sessions, a variety of scientific meetings were held with researchers and representatives from foreign institutions, who were treated to, among other activities, a gala dinner held at the Ritz Hotel with a concert of Spanish music performed by the National Orchestra in Madrid's Teatro María Guerrero.

All these events were preceded by a formal Mass of the Holy Spirit at the opening session of the meeting on 12 April, attended by ministers and senior officials in the CSIC chapel. Crucifixes and uniforms were also present at the inaugural session held in the assembly hall of the main council building where, after the dignitaries who had come from abroad were greeted, they were provided with both the relevant information about the CSIC library collection and its bibliographical services by Head Librarian Amadeo Tortajada, and the volume of CSIC publications, in numbers and sales figures, by Publications Commission President Francisco Cantera. These same publications would later be presented to Head of State Francisco Franco during the closing session held in the assembly hall, in a ritual repeated every year since the second annual plenary session in December 1941.[2]

'Upon founding the Spanish National Research Council ten years ago,' continued Ibáñez Martín in his closing session speech, 'we sought to create a corporation that would provide a refuge and incubator for the most noble cultural plans of our time'.[3] In actual fact, however, those plans – models for action drafted to guide and channel Spanish culture – were now more than a decade old. In all likelihood, their earliest manifestation dated back to 1938, when Ibáñez Martín found himself in Nationalist-held Burgos with José María Albareda, his indispensable collaborator (Malet, 2009; Sánchez Ron, 1999: 329–71). At that time, Ibáñez Martín had left behind the vicissitudes of his life in the Spanish capital in the wake of the outbreak of the Spanish Civil War where, like so many other enemies of the Second Republic, he found refuge under a foreign flag (Cervera Gil, 2006: 345–80).

However, in April 1950, amid the closing ceremony marked by pomp and ostentation, the months of war that Ibáñez Martín spent in refuge at the Turkish legation in Madrid alongside his family and a group of Requetés, seemed far away, as did the voyage on the vessel *Karadeniz* and his rushed arrival in Nationalist Spain where – in Ibáñez Martín's own words, spoken in 1937 – 'the heroic imperial, Christian meaning of the jubilant days of our greatness had caught fire' (Formentín Ibáñez, 2015: 29). While these traditional values were all associated with a conception of *Hispanidad* and a tradition of thinking that was glorified by Franco and his supporters, José Ibáñez Martín – as a secondary school teacher of geography and history, raised in a traditionalist Aragonese family in the miniscule village of Valbona near the city of Teruel – had been publicly and politically championing them since the 1920s (Castán Tobeñas, 1962: 49–56; De Balbín Lucas, 1970; Formentín Ibáñez *et al.*, 2015: 15–27; Jordana de Pozas, 1998; Juliá 2000: 275–315; Saorín Pérez and Blasco Gil, 2017).

The political convictions of Ibáñez Martín, which led him to join the Patriotic Union (*Unión Patriótica*, UP) and the National Catholic Association of Propagandists (*Asociación Católica Nacional de Propagandistas*, ACNP) during the dictatorship of Miguel Primo de Rivera (1923–30), as well as to serve as a parliamentary deputy for the Spanish Confederation of Autonomous Rights (*Confederación Española de Derechas Autónomas*, CEDA) during the Second Spanish Republic (1931–9), are well known. Indeed, these credentials made him an ideal representative on the cultural mission to South America organised by the nascent Franco regime's Delegation of Press and Propaganda in the autumn of 1937 (Martínez Barrios, 1998). During a trip brimming with conferences and events, Ibáñez Martín, in his capacity as official representative of rebel Spain, spoke at length about the values of the eternal, traditionalist and Catholic Spain in which he believed, and unwaveringly continued to advocate when vested with the power of the presidency of the CSIC and made Minister of National Education only a short time later in the new Franco state.

Indeed, as Ibáñez Martín assured the Caudillo and the other authorities gathered at the Royal Spanish Academy in December 1941, the CSIC was truly 'the splendid manifestation of the intellectual vitality of the new Spain',[4] and he himself its founder and first president. José María Albareda characterised Ibáñez Martín as 'more than President, the driving force and nerve centre of the council' (Albareda Herrera, 1942: 35), while scientist and politician Manuel Lora-Tamayo simply described him many years later as the council's 'undisputed president' (1970: 9). Be that as it may, this book does not aim to be a biography or to recover everything said about him by other chroniclers; rather it provides an enquiry into much of what was said – without neglecting the 'how', the 'where' and the 'why' – by a man who was, without a doubt, one of the most prominent authorities in the Franco regime, responsible for both reorganising education in Spain and founding the Spanish National Research Council: the President of the CSIC from 1939 to 1967; Minister of National Education from 1939 to 1951; President of the Council of State from 1951 to 1958; and Spanish Ambassador to Portugal from 1958 to 1969, Mr José Ibáñez Martín.

Chapter 2

Doing Things with Words: Epistemological Concerns

As a voice authorised by the dictatorship and the salient figure of the leading scientific institution in Spain, the words of José Ibáñez Martín are of particular interest when trying to reconstruct the official conception of science propounded by the upper echelons of the Franco regime and, thus, address the possible implications of this characterisation for the procedures involved in selecting and shaping science – and its actors, objects and places – under Franco's rule during the early to mid-part of the dictatorship. Accordingly, the following pages examine the vast body of speeches that Ibáñez Martín prepared and gave during the course of his public life. Although these speeches have been studied more or less tangentially by different authors (Amador Carretero, 1991; Antolín Hofrichter, 2018; Canales, 2015; Claret Miranda, 2006; González Blasco, 1980; Nieto-Galan, 2021; Santana de la Cruz, 2009; Santesmases Navarro de Palencia and Muñoz Ruiz, 1993; Tusell Gómez, 1984), a collective, systematic examination is still lacking. The following analysis looks at both the potentialities of what was said and the potential agency inherent in these speeches, as instruments of power, for the socio-discursive configuration of the field of Spanish science during Ibáñez Martín's time.

Any effort at the historical reconstruction of the meaning of these speeches necessarily requires beginning with a study of the context in which they were made and received. Fundamentally, these speeches were originally delivered aloud to a specific audience and in a given space for that purpose, elements that together give these speech acts both their meaning and the patterns that must be used to identify them (Goffman, 1981: 162–96). Such discourse was conveyed during solemn acts and in academic contexts, some of them embedded in the CSIC plenary sessions, specifically the closing sessions that took place during the early years at the Spanish Royal Academy, before moving to the assembly hall in the council's headquarters in 1947.[1]

The regime's official acts of this nature always received extensive media coverage, both by the press and the *NO-DO* state-controlled newsreels (Ramírez Martínez, 2011: 163–83), and were attended by the head of state and ministers, leading figures from the party, the army and the church, as well as members of the CSIC executive council, plenum, boards and institutes and academics, chair professors and cultural representatives from other countries. These contextual determinants define a place of enunciation that the empirical subject of enunciation himself, José Ibáñez Martín, described allegorically as a 'national council of culture, where all the dignitaries of Spanish knowledge converge to define – as at that thousand-year-old assembly of religious unity in Toledo – the new dogmas of Spanish science, reconstituted under the distinguished leadership of Your Excellency and placed at the service of the great ideals of the Fatherland'.[2] Thus, beyond the monopoly on power and the exercise of repression as state policy (González Calleja, 2000: 391–8), the Franco dictatorship had, by necessity, to be supported by science, education and culture, in other words, by 'the education of minds and hearts' to 'engender and produce a new Spain that', as Ibáñez Martín asserted at the solemn ceremony to inaugurate the 1942–3 academic year held in the auditorium of the Universidad Central, 'would take root in the soul of every citizen'.[3]

The intention of these speeches – characterised by party-line praise, rhetorical exaltation and the appropriation of the institutionalised word – was also to provide a specific world view for political ends. This suggests that caution should be exercised, given the difficulty entailed in fully matching the discourse itself to the reality of the social world (Barthes, 2003: 201–11). Accepting without question the official representations and categories would result in a record or log – limited in certain cases – when the true aim is to write a history that helps to explain why and how particular organisational systems, which enjoyed the protection of institutions, and certain individuals, who were given authority, became established and accepted as such. This is even more the case when working with speeches where the very *mise-en-scène* forms a substantial part of the speech act itself, constituting 'the final clothes' (Bourdieu, 2014) of allocutions that were exhorted from a position of authority, endorsed by the regime and spoken to an audience that was presumed, signalled and even invited to approach those discourses in a specific way. Moreover, these were all spoken by the inscribed voice of the subject of enunciation and the selected enunciators that the orator evoked during his speech through acts of incitement.

Furthermore, on an epistemological level, it is important to be aware that all this activity takes place during certain temporary discursive instances and at specific events to which the historian only has access through the inscription of what was said on a material support, with all that entails regarding the inherent alteration of the experience itself (Ricoeur, 2006: 38–57). Some elements with which to form a judgement have been lost, like the non-verbal actions of the empirical subject of enunciation, the prosody, the tempo of the enunciations, the gestures, how the body was positioned, and so forth. Moreover, once these speeches have been set in writing, they became part of a growing list of council publications that the CSIC Publications Office attempted to disseminate around Spain and abroad, with more or less success (López Sánchez and Fernández Gallego, 2021). With this aim in mind, many of Ibáñez Martín's speeches were not only included in the *Revista Nacional de Educación* and the Spanish National Research Council annual reports, but were also published separately (Ibáñez-Martín Mellado, 1998: 16–19). For example, as was customary at the time, Ibáñez Martín's speech to commemorate the tenth anniversary of the CSIC was printed as a paperback brochure of almost thirty pages, with a print run of 1,500 copies, by Silverio Aguirre (38 Álvarez de Castro Street, Madrid).

When working with written, transcribed and published speeches, therefore, the analysis owes a great debt to semiotic perspectives (Castañares Burcio, 1994; Eco, 1993; Wodak and Meyers, 2003), without being bound to a strictly linguistic or textualist approach (Lozano *et al.*, 1993; Van Dijk, 1996). It must take account of the enabling conditions of the speech and the multiple determinations operating in the construction of the meaning, of trying to analyse the *intentio operis* inscribed in the text (Eco, 1987). This means eschewing the fetishism of language and the pure act of verbal enunciation to make the discourse an interdependent mediator, placing importance on both what it transmits and what it expresses. Accordingly, speech acts inscribed on paper are handled as objects of study that assemble diverse referents and generate effects of reality in the text that can – and could – play a part in altering reality itself. Beyond the solipsism of a merely linguistic perspective, this entails taking into account both the forms and the content of the discourse as well as the conditions of possibility inscribed in words, times and places (Latour, 1991).

Consequently, these speeches provide a textual and discursive space that is convenient for observing the different forms of presenting and constructing the official scientific identity in Franco's Spain. Furthermore, this observation leads to a diverse range of categories that gradually shape relationships of belonging – based on religion, politics, occupation, gender – and draw boundaries between groups. These are often associated with positions in the social space and the power to grant recognition to agents of the state, in this case the president of the Spanish National Research Council, as the institutionally legitimised spokesman for groups described and represented within the institutional framework of an organisation able to represent and institutionalise them (Bourdieu, 2001: 101–30).

'Men of science,' said Ibáñez Martín, addressing Francisco Franco on 7 February 1958 at the closing session of the fourteenth plenary session of the CSIC, 'reaffirm through my words their commitment to Your Excellency as the creator of this Council and paramount driving force behind the national culture, to which they dedicate their lives, as the noblest way to serve the ideals of a better Spain'.[4] This representation also took on a performative role when those represented were interpellated based on their normative status, imposing, in turn, the authorised, legitimised and official view of science and all things properly scientific.

Chapter 3

Dismantling the Official Discourse around Science: Defining and Delimiting all Things Scientific

Since its inception, the Franco regime tried to find in science – 'this area, visibly far from all combat' – the legitimising foundations that underpinned its work.[1] Thus, in the official narrative, science soon became, metaphorically speaking, the phalanx, the infantry corps and largest army force, 'where the resilient, intact serving veterans of knowledge and culture rush to heed your clarion call to undertake the conquest of a Spanish science', as José Ibáñez Martín said in allusion to Francisco Franco's leadership at the first plenary session of the CSIC. This science, then, had to 'conquer' – using the language of war so characteristic of that time – in order to reach the objective of the complete rebirth of the Spanish scientific tradition. Ultimately, this would also entail the culmination of the spiritual revolution that was considered essential to inaugurate this new scientific era, which was to be 'the most fruitful creation of the victory and the most living essence of the glorious National Movement'.[2]

To achieve this aim, from very early on, the regime was able to count on the Spanish National Research Council (Canales, 2015 and 2019; Pastor and Sánchez Álvarez-Insúa, 1996; Sánchez Ron, 1998; Sánchez Ron, 2021; Santesmases Navarro de Palencia and Muñoz Ruiz, 1993). This new national scientific institution was founded on 24 November 1939 by the Franco regime on the rubble of its predecessor, the Board for Advanced Studies and Scientific Research (*Junta para la Ampliación de Estudios e Investigaciones Científicas*, JAE), which had been created in 1907 by Minister of Public Instruction and Fine Arts Amalio Gimeno (Caballero Garrido, 2010; Otero Carvajal and López Sánchez, 2012; Puig-Samper Mulero, 2007; Sánchez Ron, 1988;). An heir to the principles of the liberal Free Institute of Education (*Institución Libre de Enseñanza*, ILE), the work done by the JAE represented an unquestionable advancement for Spanish science that was brought to an abrupt end when the official cessation of its activities was decreed – in the midst of the Civil War – followed by its dissolution. From that time on, all its centres became affiliated with the dictatorship's new scientific institution: the CSIC.

The founder and first president of the Spanish National Research Council was Minister of National Education José Ibáñez Martín, who at all times worked in close collaboration with José María Albareda, the secretary-general of the council (Gutiérrez Ríos, 1970; Malet 2009; Santesmases Navarro de Palencia and Muñoz Ruiz, 1993). From its founding, the CSIC was the primary research organisation in Spain and governing body of Spanish science policy, displacing the Institute of Spain (*Instituto de España*, IdE), then a mere liaison of the Royal Academies, from its initial duties (Malet, 2008: 214–35). However, in the late 1950s – and in light of the unworkability of the Spanish scientific coordination model – the Scientific and Technical Research Advisory Committee (*Comisión Asesora de Investigación Científica y Técnica*, CAICYT) entered the picture. Founded in 1958, this organisation constituted the first of a series of state initiatives to provide Spain with a coherent scientific policy (Muñoz and Ornia, 1986; Nieto *et al.*, 1982). Many years later, however, when speaking about the Founding Law of the Spanish National Research Council, the president of the CSIC (1967–71), Manuel Lora-Tamayo recalled that:

> Immediately after the end of the Civil War, Spain was characterized by an atmosphere of religious and patriotic fervour . . . and it is no surprise that this atmosphere weighed on the legislators, influencing their measures; this they considered a spirit of salvation.[3]

In a grandiloquent tone consistent with the explicit objective of a legal text that aspired to impose a combination of 'the purest lessons of the universal, Catholic tradition and the demands of modernity' on the order of culture, the Spanish National Research Council was officially established in Spain 'to systematize research . . . and place scientific and technical organization at the forefront of the nation's problems'.[4] This was all under the patronage of the head of state, represented at the CSIC by Minister of National Education Ibáñez Martín, who was authorised to interpret, clarify and apply this law as he saw fit.

Thus, amid 'hymns of triumph and ration cards',[5] the CSIC reached one of its foundational goals during its first years: to dismiss the legacy of the JAE and its researchers, now seen as reminders of a past that the regime wished to eradicate using the violence and repression of the national security and public order system of the so-called New State (González Calleja, 2000: 391–8). And indeed, by 1940, nearly half the 580 university professors active before the war had died, been purged or been exiled from the country, a proportionally higher loss than that recorded in countries like Germany, Portugal and Italy (Herran and Roqué, 2013: 209). This also explains why, among the researchers working at the CSIC in 1955, only 5 per cent had published before 1940 (González Blasco and Jiménez Blanco, 1979: 131). The elimination of people, texts and names was only one of the dramatic consequences of the well-known and widely studied destruction and diaspora process of the scientific community created in Spain during the so-called Silver Age of Spanish culture, from 1898 to 1939 (Claret Miranda, 2006 and Otero Carvajal, 2006 and 2014).

In this context, it would be a mistake to take the very idea of science for granted, given that history has shown that both the concept and its limits and those of scientific work have changed over time (Barona Vilar, 1994: 11–21). Moreover, in the context of a dictatorship, interest lies not only in the categories within the scientific field, but also – and perhaps, especially – the categories taken from the field of politics that support and permeate its political power (Bourdieu, 2000a). In other words, at play are the official and even possibly ideal criteria of scientificity, as well as the collection of ideological elements involved in the processes of shaping and demarcating what science is, which, in turn, are coincident with the beliefs and perspective of the leaders of the regime. These concepts, categories and worldviews establish the outlines of what is thinkable and, therefore, what is sayable and what is feasible (Fernández Sebastián, 2009: 19). Moreover, in Spain they were all linked to traditions of thought that were not only deeply rooted in Ibáñez Martín's speeches and discourse, but also decisively involved in the development of a dictatorship that bestowed great importance on Catholics, whether in the regime's founding bureaucracies or the select circles of Opus Dei (Juliá *et al.*, 2007: 155–96). In the words of Ibáñez Martín, speaking at the event to celebrate four hundred years since the birth of Cervantes in October 1947:

> The permanent spirit of Spain. This treasure, given to us as a legacy of honour, was what we defended, flushed with patriotism, with the arms of our crusade and now, alert with promise, is maintained, without wavering, by the Franco government. For nothing or nobody that is not our own historical tradition will we abandon our position. This position . . . has centuries-old foundations, and our very raison d'être rests on it. (Ibáñez-Martín Mellado, 1998: 30)

Once again, the official documents of the CSIC provide useful material not only to learn about the regulatory framework of the council (López Sánchez, 2016), but also to observe in the textual structure the different ways of presenting and constructing a specific concept of science: the official conception. Part of the rhetorical scaffolding and the narrative of this official conception of science can already be detected in the text of the CSIC Founding Law dated 24 November 1939, which advocates for the recovery of an alleged unitary tradition of science. In his description of the text, written in 1946 for the Lucerne-based conservative Catholic periodical *Vaterland*, Swiss journalist James Schwarzenbach (1911–94) wrote, 'we find again, for the first time in the modern history of the West, the expression of the creative cultural will inspired by Catholic thought'.[6]

The Unitary Tradition and the Aspiration towards God: The Ideal Essence of the New Science

As a concept, the idea of unity has always evoked feelings with profound implications, not only metaphysically, but also in the sphere of politics and the field of scientific thinking (Galison, 2016). In fact, after the First World War, the idea of the unity of the sciences took on a modern connotation in Central Europe around a new scientific conception of the world (Carnap, 1992; Frank, 1949; Kraft, 1966). Based on the primacy of logical, empiricist and positivist analytical methods, the aspirations of the members of the Vienna Circle at that time were focused on creating a unified modern science, with a common method and language that were incompatible with theological postulates and traditional metaphysics. This unifying movement and project were developed in a series of international meetings and publications – including, most notably, the more than twenty volumes of the *Encyclopaedia of Unified Science* – but were also openly opposed to the spirit of the time, inter-war Europe, which was increasingly imbued with the theologising and traditionalist postulates that ended up driving the movement's main exponents to the world of American academics (Stadler, 2011). At a speech to mark the beginning of the 1939–40 academic year in the auditorium of the Universidad Central in Madrid in October 1939, Ibáñez Martín made his views clear in this respect: 'I begin by condemning . . . the men of cold and secular intellectualism . . . with their scientific Pharisaism, who are most responsible for the spiritual decline of our people' (Ibáñez Martín, 1939: 6). At that time – the dawn of the Franco dictatorship – this damning reference to treacherous intellectualism was almost a cliché after the proliferation of defamatory pieces during the years of the Spanish Civil War that had wielded this argument, primarily against the Free Institute of Education (ILE) and the Board for Advanced Studies and Scientific Research (JAE), the traditional targets of the criticism of the conservative and political Catholic sector of pre-war Spain (Sesma Ladrin, 2013).

It was at the Universidad Central in Madrid – where only a few months earlier, members of the Spanish University Union

(*Sindicato Español Universitario*, SEU) had set fire to pages written by Rousseau, Voltaire, Marx and Gorky (Martínez Rus, 2012: 366) – that the minister of national education held reason responsible for war crimes while, paradoxically, expressing the New State's commitment to ensure the so-called unity of science. Nonetheless, and in line with other countries like Austria and Portugal once ruled by Catholic-inspired regimes, Ibáñez Martín here was calling – as he later did during the drafting of the Founding Law of the CSIC – for a restoration of the conception 'of the classic and Christian unity of the sciences, destroyed during the eighteenth century'.[7] In other words, he was flying the flag of unified science, but in this case, as a response to the liberal scientific tradition of the Enlightenment that the authorities of the regime perceived at the time as upsetting the moral order. Moreover, in the words of Juan Zaragüeta Bengoechea (1883–1974), the vice principal of the Luis Vives Institute for Philosophy at the CSIC and doctor of theology and philosophy, this was also 'a reaction against the omniscience and omnipotence of scientific knowledge' that revealed the deficiencies of positivism, instead of the desired scientific knowledge, which was organically articulated within an integral culture (Zaragüeta Bengoechea, 1942) inspired by the fundamental principles of the New State.

To accomplish this, the official discourse of the regime turned to postulates that were directly linked to the European counterrevolutionary ideological thinkers of the late eighteenth century and a specific scientific and philosophical tradition that predated the Enlightenment. This ideological conception and view of science was synthesised in – and capitalised on – the traditionalist historical thinking of Marcelino Menéndez Pelayo (1856–1912), the intellectual reference point for the conservative, Catholic sectors in Spain (Álvarez Junco, 2001; González Cuevas, 2005; Santoveña Setién, 1994) and a figure who often appeared both in the rhetoric of Ibáñez Martín and in the naming of the various boards and institutes that comprised the Spanish National Research Council, many of which honoured the memory of the savants from that imagined Spain (Pasamar Alzuria, 1991).

'Our science today, combined with that which in past centuries defined us as a nation and as an Empire, wants above all to be Catholic'. Such was the proclamation of CSIC President Ibáñez Martín on 30 October 1940 during the reading of his keynote speech at the closing session of the first plenary meeting of the CSIC, before an audience of academic, civil, religious and military authorities, who themselves became participants in the text. Leading figures from the regime, such as the Caudillo himself, had earlier listened when Antonio de Gregorio Rocasolano, president of the Alfonso X el Sabio Board and vice president of the executive council, extolled Spain's historic collaboration in the rebirth of the sciences and scientific research, which, he said, 'would rise again in all its plenitude, organized by the Spanish National Research Council, with the assistance of God and the efforts of all'.[8] Speaking in glowing, bold terms, Rocasolano, a renowned chemist from Aragon, referred to the contributions made by that immemorial Spain to the development of the exact and natural sciences in the late fifteenth and early sixteenth centuries. This was a crucial time, the flowering of *Hispanidad* and the splendour of a universal, Catholic Spain that – in his narrative, and in line with the thinking of Menéndez Pelayo – was not the site of the scientific backwardness that others, nonetheless, accused it of.

Imbued with the same idea as Rocasolano (Ibáñez Martín, 1939 and 1940), Ibáñez Martín also viewed the prolific Menéndez Pelayo as the posthumous victor in the nineteenth-century polemic over science in Spain (Santoveña Setién, 1994: 48–68). This debate from the past was resumed and updated to articulate a new narrative that gave the official discourse of the regime a genealogy with a specific order of events and based on reference points that provided a tradition of comprehension (Goffman, 2006: 23–42). While the eighteenth century represented the period when the divorce occurred between the speculative and experimental sciences, Spain's nineteenth-century decline corresponded to the argument of subordination to foreign scientific influence in a nation that was erroneously guided by those who had allegedly forgotten the national, Catholic status of science. In the institutional discourse of the dictatorship, this forgetfulness was considered the source of a good part of the evils existing in the fatherland (Santoveña Setién, 1994: 197–243). Accordingly, on 29 January 1949 at the closing session of the ninth plenary meeting of the CSIC, Ibáñez Martín reiterated his thinking:

> The evidence of this hurtful, bitter trickery, of the lesson of the Spanish prostration during most of the nineteenth century, when an awareness of their personality lay dormant, the Spaniards, with routine neglect, abandoned the exploitation of their homeland for borrowed foreign techniques, trusting the guidance of their culture to strange ideological conceptions that denied the Christianity of Spain.[9]

Thus, the official discourse, expressed here by Ibáñez Martín, defended a purely national conception of science that was supposedly unitary and harmonious (Santana Cruz, 2009). Its programmatic premises were already sketched out at the first plenary meeting of the Spanish National Research Council. This science was indebted to a specific tradition of thought with rigorously Catholic roots, 'subjected to the supreme reason of the universe' and 'in harmony with faith ("the true Light that illuminates every man coming into the world" (John 1:9)'. However, as articulated in his narrative, this science was still presumably new, markedly Spanish and conceived as 'the effort made by intelligence to possess truth, as an aspiration towards God, as philosophical unity, as the realization of progress'. This science was underpinned by the Thomist equation, in his words: 'a science for truth and for good', which emerged from recovering a lost tradition, a line of thinking linked to the past ages, where the pinnacle of Spanish science was thought to be found.[10]

In the presence of the Caudillo during the closing session of the first plenary meeting in the assembly hall of the Royal Spanish Academy, then, Ibáñez Martín advocated 'reviving the radiance of its past scientific tradition', which would gather new momentum following the guidelines of the 'new' national science, 'a science defined according to the canons of our golden culture, but within the framework of the needs of today and submitting obediently to the Yoke of the national interest and prosperity'.[11] In its zeal to return to a lost Christian unity, this science was committed to theology as the guarantor of the scientific renaissance, understood, from an essentialist point of view, as 'the moulder of the plans laid out by the intelligent spirit, the expression of spiritual activities, the fulfilment of divine law'.[12] Ibáñez Martín continued to stress this point, for instance, during the inaugural act to mark the beginning of the 1941–2 academic year at the Universidad de Barcelona:

> This is why Spain, which never ceased to be Catholic, returns, as in the greatest moments of its Empire; in the eternal charters of Theology and the struggle against rationalist positivism, it endeavours to foster and disseminate the doctrine of that branch of science that brings man closest to knowing God.[13]

It stands to reason that, from the time of the creation of the CSIC, theology acquired a fundamental status, not only symbolically, but also materially in terms of the space carved out for the subject in and by the council. In the words of José María Albareda, the place of the CSIC itself was 'a true City of God'. In addition to the theology weeks and bible weeks held in Madrid at the beginning of each academic year and the publication of theological journals by the CSIC Francisco Suárez Institute for Theology, in 1946, the former library/auditorium of the JAE – which had been severely damaged during the Spanish Civil War – was converted into the Church of the Holy Spirit, an unequivocal symbol of the union between science and the altar, and the model for the students at the adjacent Ramiro de Maeztu National Institute for Secondary Education (Camprubí, 2014: 52–5). All this took place in the climate of scholastic thinking that was prevalent in philosophical and Church circles at the time and consistent with Thomist-Aristotelean perennial philosophy, an orthodox tendency that, while unquestionably dominant, was not monolithic (Rivera, 1978: 277–81).

In this regard, Ibáñez Martín, who could hardly be suspected of heterodoxy, always expressed his preference for the postulates of philosophers who were not strictly Thomist, such as the Jesuit

Francisco Suárez (1548–1617), whose transcendental and providentialist thinking was deeply meaningful for the CSIC president (Ibáñez Martín, 1948 and 1962). For that reason, and due to Ibáñez Martín's express wish, the spiritual patron saint of the CSIC was San Isidoro of Seville (556–636), for whom all the sciences were united and inspired by a single transcendent aim: to reach God through knowledge. Moreover, the harmony, order and unity that, according to Menéndez Pelayo, were the most distinctive values of Isidoro's work inspired Ibáñez Martín to establish the Spanish erudite as the CSIC symbol and guide in its efforts to renew the science of the fatherland. This science was to be governed by truth, towards which, Ibáñez Martín believed, human reason was naturally attracted, and guided by the supervision of theological thought and faith in the revelation of God, whom he entrusted as 'the sovereign possessor of essential, independent, intuitive, the one, infinite and infallible science', who could 'send his Holy Spirit to Spain' and 'give us the gift of true and eternal science'.[14]

Thus, order and hierarchisation became indispensable values in the scientific crusade advanced by the Franco regime against the disorders associated at the time with anti-Christian pessimism and the denationalisation of the Enlightened, liberal tradition (González Blasco, 1980: 145–52). This tradition, moreover, was affiliated with the recent Republican experience prior to the war and against what the Spanish National Research Council stood for, as 'an orphaned heir, to reunite and restore the winnowed debris of a scientific organization broken and embezzled by other hands'.[15] To that end, contrary to the 'profligacy' of 'the period prior to 1936, of fragmentary, haughty, personalist, exclusive, covered and concealing, fundamentally and deeply corrosive research',[16] Ibáñez Martín was committed to a 'creative destruction' on the part of the Franco regime, a break with the immediate past as a necessary step towards the flourishing of a sought-after 'authoritarian freedom',[17] channelled and guided by the state. Of course, according to Ibáñez Martín's narrative, the ravages of the previous experiences, now identified as heretical, cut off from Christian values and associated with a specific scientific tradition

– 'divisive and decadent positivism' – were disruptive not only to Spain, but also on an international level where, he said, they would only have served 'to cover the ruins of the ancient land of Europe'.[18] Ibáñez Martín expressed it thus in October 1940:

> Our science, the Spanish science of our Empire, which wants to most vigorously galvanize the new Spain, repudiates the Kantian thesis of absolute rationalism and is not demeaned by recognizing that man cannot come to possess all truth through continuous progress. Happy are they who apply their daily intellectual efforts to reaching God, to focus on Him as the greatest aspiration of their existence, ideas about beauty, art, philosophy and the fatherland, hand in hand with the ideals of the evangelical virtues, as a compendium of the science of life.[19]

The official rhetoric advocated the end of one cycle and the arrival of a new era, the end of the predominance of one scientific tradition considered a failure, and the emergence of another unitary, ecumenical tradition associated with the regime that comprised an episteme that viewed nature as a *mater generationis*.[20] Access to knowledge was understood to be restricted by the human senses since, as Pope Pius XII argued, in order to create science, humans had to be assisted by a higher being, the ultimate origin of all knowledge (Echarri Goicoechea, 1956: 336–49). Therefore, the official narrative argued that all proposals that 'deified human reason, holding it up as the supreme principle of all knowledge'[21] must be forgotten and rejected. For Ibáñez Martín, however, this did not diminish human intelligence; the recognition of man's limitations and the imposition of moral norms sought to perfect the scientific method by eschewing 'the voluntary, anti-scientific and anti-human amputation of our spirit'.[22]

The above words were spoken in 1943 by engineer Juan Marcilla (1886–1950), the third CSIC vice president and director of the School of Agricultural Engineers (formerly director of the Wine Research Centre of the JAE's National Foundation for Scientific Research and Reform Trials (*Fundación Nacional para Investigaciones Científicas y Ensayos de Reformas*, FNICER)). Marcilla also argued that the organic integration of the different research tasks and sciences consequently raised the level of the more modest endeavours by giving them a common goal – 'grains of sand, but grains of sand to be consolidated in a magnum opus' – by all the constituent bodies of the CSIC. Marcilla continued in the following terms at the beginning of his lecture to mark the closing session of the fourth plenary meeting of the CSIC on the topic of Spanish Possibilities for the Biological Synthesis of Proteins:

> The fundamental orders that you have given the Spanish National Research Council establish the ultimate aims of all its works done in the service of God, as the absolute and eternal Truth, and for the Fatherland, the greatest of human truths, because it integrates tradition, the present and the future, the community, the family and the individual. It is in these orders that our president, the Most Excellent Minister of National Education, has based the entire structure of the Council, unifying the ideals of quite disparate sciences in his working methods and future objectives. Walking towards these goals, which God, in His infinite mercy, has granted the grace of faith, we can see our proposals through to the very end, without fear or bias, knowing that the results cannot be, in short, contrary to the highest Supreme Truth, which is also Supreme Goodness and Supreme Beauty . . . and we write freely. But, as Ampère once said, 'Deal with the things of the world with one hand; with the other he takes hold of his Father's dress'. And I allow myself to add: 'and remembering that we are Spaniards'.[23]

Thus, the official discourse of the regime conceived of science as a synthesis between the traditional and the new, resting 'on the cornerstone of a spiritualist and renewed concept of the world', intended to endow the work of the CSIC and the new concept of science in Franco's Spain with epistemic roots and moral legitimacy. Moreover, this conception of science sought 'to ensure, along with the development of the sciences of the spirit, a concurrent and harmonic flourishing of the sciences of matter'.[24] This concept was rooted in the symbolic model of Ramon Llull's (1232–1315) *Arbor Scientiae*, or Tree of Science. For the regime,

Llull (referred to as Raimundo Lulio in Spanish and often known as Raymond Lully in the English-speaking world) was the personification of the labour of 'scientific evangelization' that the council had taken on,[25] the 'Enlightened Doctor', the 'Almogavar of thinking' and the 'mystic apologist of the unity of science'.[26] Llull's *Arbor Scientiae* became the emblem of the CSIC as the Franco regime sought to link itself to that tradition through the semiotic act of citation, perhaps tinkering with using it to strike down one of the terms of the relationship (Foucault, 1992: 23). In other words, an exercise of metonymy was at play, in which the part stood for the whole, with the cited part from tradition underpinning and replacing – hiding – the discourse of the regime which, in matter of fact, deployed this image, this allegory of the 'tree of science', adapting it and nourishing it this time with the fundamental principles of the Movement.[27]

The fatherland would give the *Arbor Scientiae* representing the CSIC 'the necessary elements to live, later returning to Spain the best fruits for the present and the promise of better ones for the future'.[28] This tree stood erect and 'points the Truth out to Heaven, that is, to God',[29] as the premise of the new science. It did not only respect the hierarchy of the sacred science, prioritising theology, which now had its place on the trunk of the new tree, but 'it yearns, as did Menéndez Pelayo, to inject new theological sap into all our cultural activities, so that science, national and thus emphatically Catholic, will above all serve the high spiritual interests of God and his Church'.[30]

Chapter 4

The International Influence: Role Models from Abroad

Against the 'Technical Spirit': Pope Pius XII and the Defence of the Catholic Cultural Renaissance

> What certainly fulfils all the aspirations that I could have as a Catholic and a Minister . . . is the letter that the august personage of the auspiciously governing Pope had the great kindness to write to me on 20 May of this year.
>
> José Ibáñez Martín, 15 December 1943, closing session of the fourth annual plenary meeting of the Spanish National Research Council[1]

The letter to which Ibáñez Martín refers here was received in response to a shipment to His Holiness Pope Pius XII of complimentary copies of texts of a historical/religious and theological nature published by the CSIC. In the missive, the pontiff congratulated the president of the council for the endeavours of such a 'meritorious Institution, called to counteract the pernicious influx into Spanish wisdom that so regrettably occurred due to sowers of a bad seed'. The pope also asked God for this work to completely 'imbue all national thinking and life', such that it would 'profoundly and sincerely reach the depths of every conscience'.[2] With this plea, Pius XII blessed not only the programme for Catholic cultural renaissance undertaken by the CSIC, but also the political orientation of the Franco regime towards the institutionalisation of the reactionary stasis of the Church.

At that particular time, CSIC Vice President José García Siñeriz was enrolled in the Vatican's Pontifical Academy of Sciences, and the cardinals of Rome, Eugène Tisserant (1884–1972), Giuseppe Pizzardo (1877–1970) and Federico Tedeschini (1873–1959), were full of praise for the council's activities. Two of these men, moreover, were quite familiar with the situation in Spain. As the secretary of the Congregation for Extraordinary Ecclesiastical Affairs, Pizzardo had sponsored Catholic Action in Spain and had guided the political ventures of the Spanish Confederation of Autonomous Rights (CEDA) since 1933. Tedeschini, in turn, served as nuncio in Madrid between 1921 and 1936, during which time he and Ibáñez Martín were united by a shared enthusiasm for the conservative project of the National Monarchist Union (*Unión Monárquica Nacional,* UNM) (Rodríguez Lago, 2017), which ultimately came to nothing amid the tragic political situation developing in Spain.

Two months after the Spanish Civil War broke out, on 14 September 1936, Pope Pius XI hastened to bless the fallen martyrs in Spain from Rome, calling them the spoils of anticlerical fury. It was during the final months of that year that then Cardinal Secretary of State at the Vatican Eugenio Pacelli came to fully understand the Catholic beliefs of both the Generalissimo and the members of the Technical State Junta, along with their explicit intention to ensure that the work of the government was permeated with Christian meaning. One year later, as Ibáñez Martín was heralding the virtues of eternal Spain against the Red barbarians while on a cultural mission to Latin America, the 'Collective Letter of the Spanish Bishops', addressed to the Catholics of the world, was already circulating abroad, gathering support for the rebel side from a majority of the episcopates. And on the very day that the victory of Franco's forces in the Spanish Civil War was declared, Cardinal Pacelli, now Pope Pius XII, congratulated Francisco Franco by telegram, urging him to lead Spain down the Catholic path that had brought greatness to the country (Casanova, 2001).

All of the above underscores the importance of considering Spain in the context of the broader international developments taking place at this time. In the same way that the Franco regime received the apostolic blessing of the Vatican, Minister of National Education and CSIC President Ibáñez Martín found an argument from authority – this time from abroad – in the controversial figure of Pope Pius XII (Friedländer, 2007) and his scientific preaching. The pope thus became a chosen enunciator, who Ibáñez Martín would cite profusely in his speeches to reinforce both his own narrative and the official discourse of the Franco regime regarding science:

> Science, which had denied the spiritual life, while it vainly thought to have acquired full freedom and autonomy by disowning God, sees itself punished today through the most humiliating bondage, having become the slave and practically automatic fulfiller of policies and orders that do not take into account whatever the rights of truth and of the human person.[3]

These words, which Pope Pius XII addressed to the world in his 1943 Christmas message, were used by Ibáñez Martín on 29 January 1949 in his own speech during the solemn closing session of the ninth plenary meeting of the CSIC as an argument from authority, to emphasise before Franco that Spanish science had not forgotten that it was only a means to a transcendent end, far, he said, from the 'obstinate materialist insipidness of nations that are hungry for bread and Spirit'.[4] However, other past political appeals once found in Ibáñez Martín's discourse – such as the recourse to the Falangist component of the regime, and the ideological affinity between Franco's Spain and Hitler's Germany – which were once considered suitable, were now strikingly absent or blurred. In 1949, of course, the Western world that emerged from the Second World War had been divided into two antagonistic blocs, and Spain, isolated internationally, was resorting to its anticommunism and Catholicism as meagre sources of legitimacy at a time, moreover, when the Vatican was calling for the incipient process towards a unified Europe to be based on the values of its common Christian heritage (Dominguez Castro and Rodríguez Lago, 2020: 306–14). To that end, Minister of Foreign Affairs

Alberto Martín Artajo, speaking at the closing session of the First Intellectual Cooperation Congress in Madrid on 12 October 1950, asserted that, 'even today, the Christian understanding constitutes the most living and working element of the cultural unity of the Old Continent. Hence this Christian understanding can and must still be the most vigorous force in the renewal of European society'.

Appealing to the 'holy doctrinal intransigence' of the regime, Martín Artajo reminded his cosmopolitan audience how Spain had victoriously shaken off the communist yoke that now blighted the West (Martín Artajo, 1950: 13–14 and 16). This was another clear example of how the Franco dictatorship had known how to read the current international circumstances in terms of religion and politics, not only sending Martín Artajo to the Ministry of Foreign Affairs, but also appointing Joaquín Ruiz-Giménez ambassador to the Holy See, fostering a rapprochement with Rome (although there had never been a separation at the doctrinal level).

Neither had Jose Ibáñez Martín at any time distanced himself from the Church. Indeed, in December 1941, he hastened to employ the words of Pope Pius XII speaking before the Pontifical Academy of Sciences to highlight the important work being done by the CSIC and Spanish science which, he said, was none other than 'to search for the solution to the problems that distress humans in the mysteries and secrets of Nature'.[5] It was also in the pope's speeches that Ibáñez Martín found the opportune argumentative basis to defend the unity of philosophical knowledge with scientific knowledge, and the unity of the sciences as an aspiration towards a harmonic synthesis of all knowledge.[6]

Although all the papal speeches were of great doctrinal value, due to their place in the papal magisterium, they did not comprise inalterable dogmatic judgements but rather provided a secure doctrine for Catholic researchers to cling to. That said, in his speeches, the pontiff affirmed that science must always be inspired by the Christian *philosophia perennis* and understand discoveries as revelations from God who, through science, allowed man – as Pius XII said in 1956 during a speech at the Seventh International Astronautical Congress – to access the 'new truths and new knowledge' that God had laid before the world. This would be the ultimate principal, the one that 'unifies all the branches of knowledge' and illuminates the nature that is the object of the work of scientists. Previously, in his 1955 speech to the Pontifical Academy of Sciences, Pius XII described these scientists as the 'authorized interpreters' of the book of nature, recommending that they meekly submit 'to all the signs of truth that manifested themselves in it'.[7]

These principles and directives of pontifical thought created a hierarchy that made God the source and destination of science, without this in any way deifying scientific findings, which always had to be framed within the ontological reality of man. However, for Pius XII, this was precisely one of the very dangers threatening the modern world: that man, engrossed in progress and the 'technical spirit' would become insensitive to divine matters (Vázquez, 1964). Hence, for example, when the pope wrote a letter that renewed his blessing of the intellectual endeavours of the CSIC in 1947,[8] he once again praised – as he had done in his first epistle to Ibáñez Martín in 1943 – the explicit promise to aspire to God and the movement towards a single, immutable truth as the principle of all legitimate scientific knowledge in Spain, reiterating the primacy of theology in the field of the sciences of the spirit, always under the protection of the Holy Ghost, 'so that His light is never missing from your intellectual work'.[9]

'Having achieved for our Science this transcendent dimension, there is no risk in ascribing its work to the physical world, to matter and life. Thus, scientific activity can be the cause of and reason for the wealth and material prosperity of peoples' (Ibáñez Martín, 1944: 17–18). The CSIC president spoke these words on 7 October 1944 in a speech given to mark the solemn inauguration of the Faculty of Sciences at the Universidad de Valencia. In this speech, he gave an account of the achievements made to that time – through the efforts of the CSIC, the university and legislation – in the process of returning Spain to the channels of Christian thought, taking the country out of the 'spiritual morass of its old academic romanticism' (Ibáñez Martín, 1944: 18). Once again, the nineteenth century in this narrative represented the time of scientific submission to foreign ways and the spiritual disorientation of the liberal state, a time of Spanish cultural recession and positivist fervour. Against these two national ills, Ibáñez Martín now, as before, appealed to scholastic philosophy and to follow papal doctrine since, as José María Albareda also argued, 'a Science that leads to scepticism, to asking, like Pilate, "What is truth?" cannot be Spanish' (Albareda Herrera, 1942: 35).

When Albareda spoke these words on 3 June 1942 during the act to celebrate his joining the Royal Academy of Natural, Physical and Exact Sciences, they formed part of a discourse loaded with patriotism and a religious sentiment that also pervaded his conception of science and research, as noted by José Casares Gil in his own comments during the act. Indeed, Albareda's speech once again contains arguments that draw not only on the premises of papal thought – the limited mind of man, science as revelation, the mystery of God and scientistic arrogance are recurrent motifs in this and other of his texts (Albareda Herrera, 1951) – but also on the time and the official rhetoric of the regime, as well as the pronouncements of those who shared the same framework and held positions that gave them power and visibility in the council. On 29 January 1949, for instance, the optical scientist José María Otero Navascués, speaking at the closing of the ninth annual plenary session of the CSIC, asserted:

> How far we are from the mechanistic theories of the nineteenth century, which made men of science arrogant in their knowledge, believing that they would be able to dominate the ultimate mysteries of the world of the Natural Sciences. Today, delving into the study of phenomena, at every point we find the traces of Divine Wisdom and we understand the many limitations of the mind of man.[10]

For his part, at the closing of the eighth plenary session of the council on 31 January 1948, Ibáñez Martín asserted that:

> The truth, by the simple fact of what it is, is one, and science, which can only be born from truth, must also be one. The Spanish National Research Council has based its work on this generous postulate

> which, glimpsed through the diaphanous minds of the Greeks, was illuminated by revelation and became a fertile foundation for Christian civilization. Only divisive and decadent positivism was able to poison European science with the divorce and discord of the speculative and experimental sciences. But the hour has passed for positivist scientism, and the Council, which experiences the certain conviction of the traditional and the legitimate perfection of the new with simultaneous and certain conviction, has based its research input on the cornerstone of the concept of a renewed, spiritualist world.[11]

For Ibáñez Martín, these premises were not only ideal, but also indispensable to channel the nation's research, the fruit 'of the new scientific conception established in Spain with the Council'.[12] As the Holy Father recognised in his message to the Pax Romana Congress in Amsterdam in 1950, it was precisely then, in the face of the advance of the disorderly applications of science and the prestige of philosophical relativism, that there was a most urgent need to appeal to an idea of science as something that follows the most fundamental principles and most essential values.[13]

Spanish physician and historian and dean of the Universidad de Madrid, Pedro Laín Entralgo (1908–2001), addressed this contemporary crisis in intellectual life, describing it as the result of the positivism, pragmaticism and historicism cultivated by nineteenth-century European thinkers (Laín Entralgo, 1949). Arguing for a return to the certainties and hierarchies of the past, Laín Entralgo outlined a definition for the archetype of the 1950s Catholic intellectual in the pages of the CSIC journal *Arbor*. In this endeavour, this well-known scholar viewed that prototype of a man dedicated to knowledge as one of many cultivators of science, highly knowledgeable about the methods to solve scientific problems. However, he said, the main difference lay in that the Catholic intellectual knew how to escape from the common illusion of attributing 'a total and absolute value to scientific truth' and omnipotence to human reason (Lain Entralgo, 1951). This was *scientia inflat* in the words of Manuel Lora-Tamayo, who was actually appealing to Saint Paul in his personal plea against the deification of human reason in modern science (Lora-Tamayo, 1953) and warned of the dangers of the 'technical spirit' and of embracing a 'technical concept of life'. This, he later said before the Royal Academy of Natural, Physical and Exact Sciences – referring to the authorised words of Pope Pius XII – 'is, without a doubt, a form of materialism, which is repugnant to our conscience' (Lora-Tamayo, 1955: 24).

Consistent with this rhetoric and line of thinking, on 31 March 1955, Ibáñez Martín reiterated at the closing session of the council's thirteenth annual plenary meeting and in front of Franco (who shortly before and upon entering the grounds had reviewed the battalion of the Ministry of the Army), that research was a national duty and research work a source of material benefits that, however, were not risk free. According to Ibáñez Martín, modern scientific research could easily fall into 'a myopic materialism, incapable of integrating the partial aspects of reality into a higher order'. This situation was be avoided at all costs, given that – once again drawing on the words of the Holy Father – 'the mind that lets itself be seduced by a conception of a life forged by this technical spirit will have a hard time finding tranquillity, serenity and the inner life necessary to be able to recognize the path that leads to the Son of God made man'.[14]

In view of the above, the CSIC president once more confirmed the council's adherence to the intellectual teachings of the Church and its policy of 'choosing people who combine a Christian education with a technical preparation' as a way to perform efficient experimental research that would not pervert the sought-after balance between spirit and matter.[15] As Ibáñez Martín would say a few years later, spiritual imperatives are non-negotiable for any political enterprise that rejects 'being inspired by the principles of historical materialism', which at a discursive level meant renouncing the cult of science to submit 'any technical achievement, however unforeseen and fantastic it may appear, to the supreme hierarchy of moral law'.[16] Here again, the words of Pope Pius XII served as a means to argue against those who erred in their understanding of technical progress, in other words, renouncing a fully Catholic perspective. As he said on 8 February

1958 during the closing of the fourteenth plenary session of the council:

> The technical progress that invades the horizons of all peoples everywhere is welcome. But let us, as the leading figures of a Catholic country such as ours, give it its true measure and meaning. Technology, for us, is at the service of the spirit and everything that inverts these terms will always represent a more or less veiled form of dangerous materialism.[17]

American Science and Creative Adaptation: The Shadows behind the Blinding Lights of the Official Discourse

> In the structure of the Spanish National Research Council, an avenue has appeared for the special development of applied research, in anticipation, without a doubt, of the current situation in the world; provisions were made in the design for a plan that the experience of the War later brought to the foreground in the world's most important countries.[18]

Thus begins the prologue to the first publication of the series on research abroad, *Investigación y Formación Científica en el Extranjero*, published in 1947 by the CSIC's Juan de la Cierva Board and printed by the Atenas typographic workshop (25 Raimundo Fernández Villaverde Street, Madrid). This almost 200-page volume on American scientific organisation, titled *Hacía una nueva organización científica de los Estados Unidos*, was prepared by the Juan de la Cierva Foreign Information Section and provided regulatory and legal information about new US scientific plans in the wake of the Second World War. Accordingly, the publication also included the first steps taken towards founding the US National Defence Research Committee (NDRC) in June 1940, followed by the Office of Scientific Research and Development (OSRD), which absorbed the former body one year later. Both institutions were directed by the American engineer Vannevar Bush (1890–1974), the man who advised President Franklin D Roosevelt (1882–1945) on 17 November 1944 to organise a national scientific research programme that would pursue, finance and expand the achievements made by the Americans during the still ongoing conflict.

Vannevar Bush was also the author of the famous report that emerged from the Second World War, *Science: The Endless Frontier* (1945), considered the emblematic starting shot for modern scientific policy (Dennis 2004; Sanz Menéndez, 1997: 71–9). This text represented a robust defence of research in science – both basic and applied – as well as its autonomy and state funding, all of which reflected the spirit of the time when it was written. However, conditions that were initially viewed as exceptional during a period of war came to constitute the legitimising foundations for the status of modern science and subsequent scientific policy during the Cold War, both in the United States and in most industrialised Western nations (Krige and Pestre, 1997). Indeed, the general outlines of this action plan had already been set out in the first major speech given by President Harry S. Truman (1884–1972) to the House of Representatives in September 1945 in which, along with a series of proposals related to improving the economy, education and public employment, he emphasised the responsibility of the government to boost research in science, technology and medicine, also proposing the establishment of a federal research agency for the United States that, after years of litigation, was launched in May 1950 as the National Science Foundation (NSF) (Lee Kleinman, 1995).

Back in Franco's Spain, in December 1945, Ibáñez Martín presented to the authorities at the closing of the sixth plenary session of the CSIC a point-by-point listing of the main ideas in President Truman's speech:[19]

> According to the President of North America [*sic*], 'Progress in scientific research and development is an indispensable condition to the future welfare and security of the nation. The events of the past few years are proof of what science can do'. He proposed 'the

> establishment of a single federal research agency which would govern research projects, promote and support research in public health, provide financial assistance in the form of scholarships and grants for young people of proved scientific ability, finance the diverse scientific activities now conducted by the several official departments and agencies'. He later addressed coordinating research with an emphasis on freedom, using these precise words: 'Scientific knowledge and scientific research are a complex and interrelated structure. Technological advances in one field may have great significance for another apparently unrelated. Accordingly, I urge upon the Congress the desirability of centralizing these functions in a single agency. Although science can be coordinated and encouraged, it cannot be dictated to or regimented. Science cannot progress unless grounded in the free intelligence of the scientist. I stress the fact that the federal research agency here proposed should in no way impair that freedom'.[20]

Specifically, the extract that the CSIC president reproduced in his speech translated point 12 on 'Research' from the twenty-one-point programme for American reconversion, using it as an argument from authority – despite the Democratic president's well-known personal aversion towards the Spanish regime (Edwards, 1999: 42–4) – to provide support for the activities being carried out in Spain by the CSIC and its scientists.

Ibáñez Martín continued: 'Although the size varies, the problems are the same everywhere', as was the desire to find solutions, which always involved structuring the research and maximising its benefits by uniting scientific activities under a single governing body. Accordingly, the arguments in his speech validated the CSIC as this national centre that, reflecting international developments, would direct scientific research work in Spain and provide financial assistance in the form of grants and awards for scientific activity, financing all these endeavours with the Caudillo, as head of the Franco state, embodying the role of the patron of science. Ibáñez Martín, however, did not stop there. By making the words of the American president his own, he indicated that science could and should be coordinated and fomented without having to jeopardise freedom. Once again, what Ibáñez Martín truly wanted at the time was to reaffirm the message that the centralisation and institutionalisation of science in Spain would in no way restrict 'this much needed freedom' (Ibáñez Martín, 1950: 10).

Therefore, the regime chose to promote science supported by funds and institutions that were dependent on the state; thus, partially ceding its authority to the scientific community in exchange for scientific and technological products of national benefit. This instituted a *de facto* strict 'social contract for science' (Guston, 2006; Jasanoff, 2004) between the scientific community and the Franco regime, in which the latter was responsible for looking after the interests of research through forms of patronage and sponsorship developed by the state, funding studies and providing official channels to publish the results, as well as granting professional recognition and honours to experts in the field. In return, the Spanish scientific community was responsible for obtaining real knowledge, always bound by not only 'the strict method and technique imposed by science', but also the interests of national prestige, 'in industrious service to Spain under the Caudillo'.[21] Once again, all this was mixed with the rhetoric of 'free science', the ultimate guarantee of the independence and autonomy of scientific research, thus eluding any outside influence and equipped, it was said, with everything necessary to develop a kind of 'scientific-certified' knowledge in Spain (Blanco and Iranzo 2000: 90) under the shade of the CSIC tree.

The recurring argument of non-interference reached its peak in a tempered speech given by José Ibáñez Martín to the representatives of institutions from Europe and the Americas on the occasion of the tenth anniversary of the founding of the Spanish National Research Council. In his discourse of the time, the former revanchist and self-absorbed Spain was being replaced by a 'Spain of springtime and fragrant peace'. This was the period of the consecration of the 'Third Phase of Spanish Science', the guarantor, according to Ibáñez Martín, of 'the permanent and universal unity of Science, whose passionate and constructive work contradicts the circumstantial historical discrimination of winners and losers'.[22] Thus, in the early 1950s, and from a perspective – that of the 'modest, valuable scholars' who made up the CSIC – rooted in the tradition of the polygraph Marcelino Menéndez y Pelayo, Ibáñez Martín denied the reality of the exile and purge that he had actually justified years earlier. Moreover, this took place at a time when a new certainty was appearing in Spain: the scientific exodus.[23] This was an inescapably uncomfortable truth, and as argued by Gregorio Marañón (1887–1960) in February 1958, it would force many young Spanish scientists to 'emigrate to other lands to serve the religion of truth'.[24]

Nonetheless, in April 1952, then Minister of National Education Joaquín Ruiz-Giménez argued at the closing session of the twelfth plenary meeting of the CSIC that: 'The creation of our Spanish National Research Council in Spain in 1939 is another confirmation, a splendid confirmation, of how keenly Your Excellency knew how to intuit the best paths of historical progress at the correct time.'[25] Ruiz-Giménez, then, was praising the Caudillo for anticipating – in the constitution of the council – a trend in which states were established as the protectors of free science, as in France (*Centre National de la Recherche* CNR; 1945), Italy (*Instituto Nazionale delle Ricerche*, INR; 1945), Germany (*Deutscher Forschungsrat*, DFR; 1949) and the United States (National Science Foundation, NSF; 1950). However, according to Ruiz-Giménez, the state could sully this scientific freedom by exercising both 'rigorous abstentionism' – a position he called anachronistic and characteristic of 'radical liberalism' – and a 'totalitarian absorption', such as that characteristic of countries under the Marxist yoke. Instead, Ruiz-Giménez called for every state to adopt, like Spain, a 'truly Christian' position towards science; in other words, to 'operate as a stimulus, complement and coordination',[26] which, it was supposed, would not restrict the legitimate freedom of the researcher.

However, it was actually the very lack of freedoms in the Franco regime that condemned Spain to international ostracism, something ratified by the withdrawal of accredited plenipotentiary ministers and ambassadors from Madrid by the member states of the United Nations beginning in December 1946. Nonetheless, it is also true that at that very time, the international effects of what became known as the Truman Doctrine were conducive to the gradual international recovery of a Spain that was destitute and lacking liquidity (Portero Rodríguez, 1989). With the world now divided into two large, irreconcilable blocs, Spanish diplomatic efforts focused on currying favour with the United States and its anticommunist military policies, especially after the events of the Korean War (1950–3) led the North American power to set aside its reticence and sign a major pact with Spain in 1953 (Viñas, 1981 and 2003b). Moreover, the agreements in this pact would have a significant impact on Spanish security and foreign policy in the long term (Viñas, 2003b: 83–108).

It was not only the United States that had to conquer its reticence. As soon as the American hegemony in Europe became clear, the Franco regime did not hesitate to change its initial anti-American position towards one that was much more inclined to reach consensuses. In the particular case of the CSIC, for example, during the 1940s, exchanges of publications with the CSIC centres were resumed. Along with scientific and technical books and periodicals, research grants for doctors, biologists, physicists, chemists and engineers selected from the regime's scientific institutions reactivated Spain's intercultural exchange circuits (albeit with the inherent problems of credit and currency), while the United States attracted the regime's scientific elites and leaders to its cause (Delgado Gómez-Escalonilla, 2003; Delgado Gómez-Escalonilla and Rodríguez-Jiménez, 2018).

Political and strategic interests were also behind the recurring quotations and references to the United States in Ibáñez Martín's speeches, which he used as an argument from authority. The image of North American science in his discourse, however, was fabricated – and adapted – according to the orator's objectives (García Naharro, 2017b). For example, with the United States positioned as the centre point and model of scientific leadership, Ibáñez Martín definitively dropped his opposition to all things foreign in order to incorporate the 'creative adaptation' of foreign knowledge into his narrative in pursuit of increased return, committing to the 'social contract for science' and, with increasing insistence, to private capital.[27]

Clearly, like that of the regime itself, José Ibáñez Martín's discourse over time was not immutable. In fact, in some ways, his rhetoric had to gradually be tempered and adapt to a future that was increasingly more concerned about profitability and efficiency in a Western world that was heading down a clearly technocratic path, as the Franco regime itself would do during the 1960s (González Cuevas, 2007; Zaratiegui, 2019). For that reason, understanding the meaning of what was said and duly calibrating its pragmatic significance requires examining the role played by references to American science in his discourse, which now served as a mirror in which the science of the Franco dictatorship wished to see itself. However, when perceiving not only the author's subjective intention, but also the objective meaning of his text, the context cannot be overlooked (Juliá, 2004:15). Again, references to the United States formed part of the framework of the speeches made by the CSIC president to the representatives from the ruling circles, and were used as a catalyst for an ideal process of transformation and improvement in a country that wanted to reach – or at least near – the achievements of a foreign country that was positioned, discursively, as the centre and model of scientific leadership. In doing so, however, Spain then became less prominent in the narrative, while new actors came into play: foreign research centres that, surprisingly, no longer posed a danger, as Ibáñez Martín acknowledged in April 1952, to 'our unwavering unity', committed to 'maintaining our own lofty ambitions', as the country 'opens up and grows in strength amidst an unlimited horizon of collaborations'.[28]

At that time, regime politician Joaquín Ruíz-Giménez called for an intensification of these international contacts, highlighting as the achievements made in this area both 'the unanimous admission' of the CSIC's Juan de la Cierva Board Institute for the Rationalization of Work into the International Organization for Standardization (ISO) in Geneva in 1951 and the incorporation of the CSIC into the International Council of Scientific Unions.[29] These admissions into international agencies, along with past and ongoing invitations to foreign scientists and the dispatch of council members to work and study abroad, must be understood as ways to break the country's isolation and re-establish ties with the foreign scientific community, while at the same time operating – or potentially operating – politically as a tool to assist the progressive international acceptance of the regime (Camprubí, 2014; Nieto-Galan, 2015; Santesmases Navarro de Palencia, 2008).

At the same time, it was no less important that the discourse of both Ibáñez Martín and the Franco regime itself increasingly stressed profitability and technical progress in their legitimisation of science. Although the link in the argument between technical development and national and economic progress – without necessarily communing with the so-called technical spirit of the age – intensified over the years and due to developments in neighbouring countries, it was very much on the minds of the CSIC members from the earliest years, who used it as a weapon against the scientific administration of the Second Republic.[30] In this respect, the predominance of the Juan de la Cierva Board with regard to the CSIC's expenses and investments serves as a palpable example of the active role played by applied science within the self-sufficiency economic model advocated by Juan Antonio Suanzes (1891–1977), the president of both the National Institute of Industry (*Instituto Nacional de Industria*, INI; 1941–63) and the Juan de la Cierva Board (1942–63) (López García, 1994). However, when Suanzes spoke at the twelfth plenary session of the council, he expressly asked the attendees to provide more funding and other means to pursue the work of his board:

> In the past, we have asked for understanding or collaboration, support or faith; today, specifically, we must make a demand that our Superiors will be able to interpret and translate into reality better than we can, and if they consider it suitable. Money. Means so that our work can advance and bear fruit. Approximately fifty-five million pesetas is spent each year by our Board on technical research. This is what a single foreign Company, and not a very large one, invests in similar work during the same amount of time. Our immediate aim must be to at least double that amount as soon as possible. If we were to receive this today, it would be quickly absorbed and there would be an immediate return from the Institutes for Fats,

> Electronics, Research into Fisheries, the Rationalization of Work and Technical Research, in Barcelona, there is a lack of sufficient allocations for the creation of institutes of Wine, Preserves and Textile Research, which are absolutely necessary. The payback would be immediate. If we do not achieve this, we will have made a very serious mistake, and such mistakes are very costly.[31]

This need for a greater financial effort on the part of the state was endorsed by then Minister of National Education Ruiz-Giménez, who extended it to major industrial companies, large landowners and the financial sector. On 24 April 1952, during the solemn closing session attended by Generalissimo Franco, ambassadors and foreign ministers, as well as the upper echelons of the Church and Spanish state, Ruiz-Giménez spoke 'with the seriousness imposed by the respect due to You' about unreached goals, researchers without 'financial self-sufficiency', about increasing the financial resources set aside for Spanish science, not only related to material constructions, but also to staff salaries and the supply of scientific material.[32] Unquestionably, this clashed with the traditional celebratory narrative of Ibáñez Martín, whom Ruiz-Giménez had replaced the year before as Minister of Education. On 31 March 1955 during the closing of the council's thirteenth plenary session, Ibáñez Martín still praised:

> Suitable buildings, specialist libraries, modern scientific material, experimental centres, laboratories, unique research sectors conquered in the vast arid zones that materialize in the Spanish scientific context, publications, patents, new technologies for practical applications, vital connections with Ministries and Industry, international relationships and prestige, incentive and an example for other countries, but above all, researchers, collaborators, fellows, interns, research assistants; in all these sections we can present to you, Your Excellency, developments that are effective and increasing.[33]

However, although these words spoken by the CSIC president may have traditionally closed his annual speech, they were then accompanied by a plea in support of the financial value of research, loaded with references to the 'great powers' with a research tradition. This was a demand of the times. At this juncture, the previous requests for greater state funding for science made by Minister of National Education Ruiz-Giménez had also been partially compensated. As he himself noted in March 1955, 'The state budget for this Council, which was some 66 million pesetas in the 1951 Budget, rose to 104 million pesetas in the 1952–53 biennium and in the current budget is roughly 108 million pesetas'. It was necessary, according to the minister, to add to these figures both the subsidies from various ministerial departments, like public works, the army, the marines and commerce given to specific CSIC institutes, which he put at some 30 million pesetas a year, and the nearly 36 million pesetas for fiscal years 1954 and 1955 provided by collaborating companies and industries. Altogether, however, this only represented 0.03 per cent of the national revenue, which was what, in the words of the minister, Spain was dedicating to scientific research at that time.[34]

Investment in research in 1959 only accounted for 0.1 per cent of the national revenue, an amount that was clearly insufficient for the CSIC researchers. Even Franco himself remarked, 'that strikes me as a rather low percentage'[35] when speaking to the CSIC Science Division, while Dr Rius of the Institute of Physics and Chemistry, in turn, told him that 'the most overwhelming problem' at the time was 'paying people, to the extent that the Rocasolano Institute can exist with the assistance provided by the Americans'. However, after telling them that it was not the time to be able to give research everything it needed, General Franco did not hesitate to remind them that research required a strong vocational component, especially under the circumstances of the time:

> The Caudillo stated that Private Corporations had perhaps lost their perspective in awarding salaries to their officials, which were far above those of the State, and that this was a great disadvantage, especially in the case of research, which requires a strong vocation, because a regular doctor in Madrid makes enough money to live

> very well, but if he leaves his clinic to dedicate himself to research, he has to live without.[36]

In a letter written in May 1959 to CSIC Vice-President José Royo, Ibáñez Martín said that 'It is absolutely right for the Caudillo to know that research is very poorly paid at this time . . . We need to exert pressure, appropriately but efficiently, to improve the appropriations for our service'.[37] Now Spanish ambassador in Lisbon, Ibáñez Martín was expressing some concerns that had been absent from his official speeches and discourse before. There is no doubt that this letter – an example of ordinary, subjective and singular writing, halfway between public and private – created a distinct textual space, built on trust and the supposed confidentiality characteristic of the medium. Now it is possible to see how many of the silences in the official discourse came to light when it was confronted by practice and the everyday reality at that time. Public images and presences that appeared to be robust when played to the gallery can become flimsy realities when examined in the right light. Perhaps for that reason, after twenty years of sowing and nurturing the tree of the CSIC, its trunk, which also appeared to be strong and solid, really hid a more complex reality.

'It makes no difference to me whether they gain (?) [*sic*] a chaired professorship or go into private professional practice. Indirectly, but efficiently they are pruning . . . the branches of the Council that could produce more fruit!' These words were written in February 1960 by Alberto Sols (1917–89), then the president of the Association of CSIC Researchers and Collaborators and adjunct director of research at the Science Division. He resigned from those two positions that same month, disillusioned by the consequences of the ambiguous text of the 1959 reform decree for the CSIC's researchers and collaborators. In a letter to Ibáñez Martín, he described this staff, their 'distressing financial hardship' – despite wage increases, which were nonetheless 'indiscriminate and insufficient' – now combined with the temporary nature of their contracts (four years to be renewed if the institute or centre considered it appropriate). He added that what could have been a 'just weapon against neglect' had become – as 'most of us feared and felt' – an 'unjust weapon to keep the Researchers reliant on personal favour'[38] if they wished to continue their scientific work at the 'highest' national institution under Franco: the CSIC:

> Of the first generations of Fellows who, given a home by the Council, began to do research, a few Researchers are now already almost 40 years old, in other words grown men. In addition to age, some Researchers have acquired international prestige that overshadows, or tends to, some of the Directors. It relieves them that the best Researchers are stifled. And the possibility that they may leave the Council is no problem: it is easy to fill the laboratories with new Fellows and 'Temporary Workers'. It suits them to establish a comfortable mediocrity . . . This reminds me of the conclusion that I reached years ago, that there are two ways to excel professionally in any given society: to climb higher or to stop others from climbing.[39]

Part II

Actors

Chapter 5

The Figure of the Scientist: Shaping the *Vir Modestus*, Subverting Women

Writing in the late 1980s, French cultural historian Roger Chartier asserted that no practice or structure exists that is not produced by the contradictory and conflicting representations by which individuals and groups make sense of their world (Chartier, 1992: 49). Guided by that insight, the second section of this book examines the influence of the official concept of science and the shared frames of reference explored in the preceding section on the scientific field in Franco's Spain. However, this in no way means that the focus is going to be limited to external factors of some nature; rather it will extend to how the representations utilised and the social processes were in fact involved in the constitution and demarcation of the institutional sphere (Woolgar, 1991), in this case, the reality of the Spanish National Research Council and, particularly, the branches of technical research which, according to the narrative of José Ibáñez Martín, had been underserved until the creation of the CSIC.

While the analysis of these various rhetorical strategies and the recourse to argument in the first part of the book demonstrated how the official definition of science in Franco's Spain was imbued with political and ideological interests, the second section focuses on how other elements came into play in determining what comprises the field, resulting in an essentialist conception of science and introducing any number of keys to delve into the relationships between science and gender. Although the Franco dictatorship from the very beginning established virility as an intrinsic feature of the New Spain (Box, 2019), the regime also knew to give place of pride to another male model, the male/subject/knower, or in other words, the scientist and erudite. This *vir modestus*, a scholarly modest man, was viewed as a legitimate, authorised male ventriloquist of the world of nature, equipped with certain virtues characteristic of scientific asceticism, like moderation, discipline and selfless work. Moreover, through semiotic and material articulations, this professional ethos would also attract specific objects and places (Haraway, 2004).

'Over the living face of the Earth, everything turns and is produced in man' – said Ibáñez Martín to the CSIC senior staff in April 1950, adding 'and the search for scientific truth is no exception'.[1] Francisco Franco himself specified: 'men of the laboratory and library, select spirits',[2] making patently clear once again the distinctive feature of some men – he consciously employed the masculine gender – who were associated with specific, defined spaces in accordance with their chosen and select nature. Understood thus, this nature ultimately established a presumably unquestionable limit between them and other mortals (Latour, 2008: 51–7). It is little wonder that this was also one of the ways of ensuring – that is, making certain, complete and stable – the very definition of this group of 'men' of science at the council, the epitome of 'a true aristocracy of the spirit' (Ibáñez Martín, 1942).

Dating back to the 1940s, the official discourse as expressed in speeches revealed a specific model of the scientist, understood as a member of the aristocratic class, as espoused by Max Scheler, one of the free, contemplative men who would be part of the development of knowledge (Lamo de Espinosa, 2002). What Ibáñez Martín described as 'this aristocracy chosen from the men of tomorrow', were educated in scientific circles and in their hands the 'guiding principles of the future' would be placed. This class of men – again, using the masculine gender and, thus, identifying the figure of the scientist and erudite with the male sex – was chosen and identified as the new Spanish intellectual, 'those of you purified in the crucible of the red revolution and the bloody war'. These scientists were also the target audience[3] for Ibáñez Martín's speech during the first plenary meeting of the CSIC, when he charged them with the mission of using science to 'forge the national spirit' of Spanish youth, a rhetorically idealised youth to whom they, as 'organic intellectuals', had to convey 'the solid virtues required to create an authentically Spanish science that can secure the grandeur of the nation'.[4]

Returning to this subject during his speech to inaugurate academic year 1942–3 at the Universidad Central, Ibáñez Martín spoke of a science 'that bows before what is true and good, conceived as service to God and the Fatherland, demanded by the State for the common good of its material and spiritual needs' (Ibáñez Martín, 1942: 21–2). In that solemn academic context, the Minister of National Education thus appealed to the value of effort, education and intellectual development in a speech also packed with calls for the re-Christianisation of Spanish youth, praise in honour of the volunteer Blue Division fighting on behalf of the German Army (which he described as 'the vanguard of Spain against the negative forces of Asiatic atheism'), and the duty of the 'new intellectual aristocracy' to 'restore the brilliance of Spanish science under penalty of crime against the state' (Ibáñez Martín, 1942: 18–19 and 22).

These were the immediate post-Civil War years, characterised by the utopia of full national self-sufficiency and reactionary, military language, years when Ibáñez Martín insisted from his institutional position during the first meeting of the CSIC, that 'the stupidly happy time when scientific activity was a free and voluntary contribution' had come to an end.[5] Over the years, the discourse of these arguments – or at least some of them – became nuanced. However, it was at this time of rhetorical and practical fury that the official narrative of the regime began to outline an image that would accompany its research work for many years. This image was characterised by obscure, sober and silent work that resulted from a 'long and prolonged invisible effort', the antithesis of the visible exhibition that always, in the words of Ibáñez Martín, 'was an assault on the very essence of research'.[6]

Therefore, at the discursive level, being obscure and internal, costly, slow and prolonged over time became a rhetorical part of the epistemological decorum that needed to characterise scientific research, as an activity charged with 'discovering the deep, internal nerves that tie members to causes', where the exterior, according to Ibáñez Martín, is only 'a consequence of inner workings'.[7] Opposing directions were established: inside versus outside; science versus doxa. However, this did not stop the CSIC president from arguing that this same scientific work entailed 'looking up continuously and energetically, as the way to be productive and prolific', as opposed to 'looking down comfortably and deflated, spending life in judgement'.[8] Interpreted now from a Christian perspective, 'up' and 'down' refer both to the aspiration towards divine perfection and to the descent into the most base and purely human passions. Again, these narrative recourses not only serve a specific conception of knowledge, but also the construction of a moral model for scientists to follow.

At the same time, the official narrative provides a textual and discursive space to determine how and with what rhetoric both the figure of the scientist – the *vir modestus* championed in the official discourse of the Franco regime – and science itself were protected and defended, positioning the two beyond all doubt, something essentially separate and distinct from the rest. First, in this narrative, the figure of the scientist was constructed on the premise of solitary, self-reflective work, far from the clamour of society. These assumptions also guaranteed the value of the knowledge acquired, in the private context of discovery (Shapin, 2010), and that Ibáñez Martín always associated with the expert, the tranquil semi-darkness of the laboratory and the quiet enclosure of the library.

On the nature of this knowledge and research work, priest and philosopher Raimon Panikkar (1918–2010) wrote in 1942 in the *Revista de Filosofía* published by the CSIC Luis Vives Institute for Philosophy that 'knowing is discovering' and that 'researching means following the vestiges left in its wake by something existing, something real'.[9] As he himself recognised, behind his words lay a philosophical concept that was also used by José María Albareda, who cited Panikkar as an argument from authority years later in his publication on scientific research, *Consideraciones sobre la Investigación Científica* (1951), when he defined research as the act of 'delving, developing, searching for new acquisitions' that 'the limited mind of man, positioned in divine Creation' longed to find (Albareda Herrera, 1951: 15). However, Albareda (the secretary-general of the CSIC who Ibáñez Martín described as the archetypal researcher and priest)[10] was really merely recovering the definition that he himself had given in the summer of 1942 to an audience of the Royal Academy of Natural, Physical and Exact Sciences in his speech on the formative value of research. During that lecture, Alborada also argued that science was a kind of 'natural revelation' that demanded of the researcher scientific integrity, industriousness and sincerity, in addition to closing himself off to many suggestions in a disciplined manner (Albareda Herrera, 1942). To that end, Ibáñez Martín also advised 'eschewing appearances and minimal effort, because research is an internal activity and if listlessness and weakness are fatal for all human activity, for an internal activity they are completely annihilating'.[11]

Thus, classifying dichotomies and representations shaped a sort of scientific asceticism – connected to sufficient retribution, but always far removed from excess – that ultimately also shaped the professional ethos of the austere staff of the CSIC. These features and behaviours, however, should not be understood as given; rather, to a large extent, they were shaped through the scientific practices and conceptions themselves (Golinsky 2005: 103–32), as well as through the social construction associated with the figure of the scientist and his knowledge as true and reliable (Shapin 1995). Moreover, this scientific ethos was equipped in this case with modes of action, a crystallised form of organisation and means of regulating the scientific community (Shapin and Schaffer, 2005: 44), whose values were heavily loaded with Christian rhetoric: humility, constriction, asceticism, personal sacrifice. These would connect the scientists to nature, the 'nourishing mother' whom they were to respect and treat objectively, without any intervention on the part of the observer. This was accentuated by the theological claims of the Franco regime, which positioned them idyllically, on the borderline between objectivity and their pursuit of the discipline of the saints, which would bring them within reach of the perspective of the angels (Daston and Galison 1992: 81–128), by subjecting themselves – always in compliance with the official rhetoric – to the recognition of 'the Truth', which is God, from whose path one could not stray.

'The truth presented to the scientist is a truth that already exists, and all his efforts, however enormous that may be, will only succeed in perceiving it, never in creating it'. Thus did the chemist, and recently appointed Minister of Education and Science Manuel Lora-Tamayo exhort his audience at the Ateneo de Sevilla in 1965 (Lora-Tamayo, 1969: 29). With this, the former intern at the Board for Advanced Studies and Scientific Research (JAE) called for the submission of each researcher to scientific truth as the ultimate objective and as an exercise in responsibility, humility and generosity on their part. Once again, this essentialist and transcendental conception of science and all things scientific is wielded, the passive idea of science as the discovery of the secrets of Divine Wisdom that pre-exist human beings, but is actually underpinned by practices and discourses like these (Woolgar, 1991: 32). In this case, they were deployed by important figures from the scientific institutions of the regime and its ideological models.

At the same time, all these so-called virtues associated with science and managed by the official rhetoric were, in turn, historically constructed around the male gender. Attributes that appeared to be grounded in reality had been, of course, constructed around the original opposition between male and female, which led to the imposition of an entire system of androcentric classification in the division of values of excellence, as both neutral and natural: rational / emotional; rational / irrational; objective / subjective; universal and impartial / individual and personal; culture / nature; large / small; up / down; public / private and so forth (Bourdieu, 2007: 107–28; Howkesworth, 1989; Lloyd, 1984). These classification systems and guidelines for action, in turn, established equivalences between the physical space and the social sphere, thus facilitating the entrance of those who adopted these recognised and recognisable forms in the scientific field. Accordingly, definitions that were accepted as universal really upheld a particular, sexually connotated – and even legally sanctioned – way of being 'made to measure for men' (Bourdieu, 2007: 75–83). This was particularly true in Franco's Spain, where gender education and socialisation were segregated by law.

While God had reserved 'the suffering of childbirth, the labor of nursing and early upbringing of children' for women (Pius XII, 1945: 291), National-Catholic Spain mandated her subjugation to the home. An entire discourse of femininity was articulated around the household and regulated everything from women's attire to their consciences, always in accordance with the elevated mission of the spiritual and social order that corresponds to the female sex in Christian life, making her responsible for defending traditional family values (Roca i Girona, 1996: 81). These values were in open opposition to those of the emancipated woman, the result, it was said at the time, of so-called masculinising feminism, which created an 'arid intellectual product' (Di Febo, 2003: 35 and 28) and contradicted the authentically female values that viewed women as wives and mothers incarnate, and were taught in schools, normalised on the streets and reinforced by the law. Women were always considered non-autonomous, dependent subjects (Gallego Méndez, 1999: 214–15), guarantors of the transmission of values that centred around motherhood, the household and caretaking, due to the role they had to play in the family. This 'angel in the house' was presented as the ideal of womanhood sustained – and endorsed – by institutions like the family, the Church, schools and the state itself, all of which reinforced the male domination that permeated the other government practices related to security, indoctrination and social control during the different stages of the Franco period (Cayuela, 2014).

Not only regulations and decrees, but also socially shared perceptions were promoted by the dictatorship and used to make it difficult for women to feel comfortable even in supposedly neutral situations like those related to high culture, where the common sense of the time could lead a woman holding a position in the sciences to be branded a bad mother and wife with total impunity (Alcalá Cortijo and Magallón Portolés, 2008: 160). Accordingly, women were not always regarded favourably in the Spanish National Research Council. One such example from 1959 is Carmen Conde, whose status at the CSIC led Ibáñez Martín to insist, in writing, that she 'be treated like the other civil servants, in accordance with their terms' and that this be resolved swiftly.[12] María Egües, a physicist in the CSIC Institute for Optics, recalled that José Casares, a member of the JAE before becoming the first director of the Institute for Physics and Chemistry at the CSIC, among other positions (Suay Matallana, 2012), did not want women in his institute, but 'that he tolerated Piedad [de la Cierva], because she had worked there for a long time and he knew her family' (Moya de Guerra, 2002: 564).

Other women encountered a similar fate. After receiving her first degree in chemistry, María Josefa Molera was rejected as a doctoral candidate by Julio Casares Gil for being a woman; however, she went on to become a pioneer in Spanish chromatography, council fellow and, years later, a research professor at the CSIC. The biochemist Alberto Sols recalled that when Margarita Salas, later a renowned molecular biologist, first came to his laboratory to ask for a doctoral position, he thought, 'Bah, a girl. I'll give her a topic that isn't very interesting, so that if nothing comes of it, it won't matter'. Salas herself said that his reaction 'might have reflected the mentality that existed in 1961 regarding what could be expected of scientific work done by a woman' (Salas Falgueras, 2011: 20).

Unlike her experience as a student of Chemical Sciences in Madrid (1955–60), where Salas did not feel that she was discriminated against for being a woman, with Alberto Sols – her thesis director and 'the best biochemist in Spain at the time' – Salas never came to feel like an equal. Even during the 1960s and early 1970s, having studied and conducted research under both Sols and Severo Ochoa (1905–93), Salas continued to be viewed in Spain as 'Eladio Viñuela's wife'. However, although her marriage to an outstanding scientist (also a pioneer in the field of molecular biology) may have obscured her contribution as a researcher, it also provided convenient social coverage to allow her to engage in professional interactions that might not have otherwise been considered appropriate for a woman in a world of men (Santesmases Navarro de Palencia, 2000: 108–14). Undoubtedly, in Franco's

Spain, when paid work outside the home by a married woman was viewed as a threat to the stability of the nuclear family and an affront to the man's role (Roca i Girona, 2005: 87–91), for women in science, having a partner and living in an environment where an interest in knowledge was nourished made things easier.

Proximity to (male) scientists equipped these women with essential social capital, without which they would have found it difficult to gain visibility and access to opportunities at a time, and in an environment, dominated by strong sexist prejudice. Consequently – and following in the footsteps of recent examinations of the relationship between science and gender in Spain (Barral Morán, Magallón Portolés, Miqueo Miqueo and Sánchez González [coords], 1999; Magallón Portolés, 2004; Ortíz Gómez and Becerra Conde, 1996; Pérez Sedeño and Alcalá Cortijo, 2001; Romero de Pablos and Santesmases Navarro de Palencia, 2017; Santesmases Navarro de Palencia, 2000) – the next chapters delve into the histories of the women who achieved legitimacy and a sufficiently independent status to be able to choose to be 'modest witnesses' (Haraway, 2004) during the Franco period. Methodologically speaking, this exploration is done again using particular printed products, which are interrogated here not only as texts, but as products in circulation that convey information of interest beyond their content. With and through these texts, it will be possible to challenge the way in which specific, apparently intangible knowledge and perceptions nonetheless acquired a material form and presence (Howsam, 2016). And now to begin.

Chapter 6

Women Scientists on Paper: Science, Gender and Power in Franco's Spain

'Science and Technology need their men, their selfless, dedicated men.'[1] Thus begins the introduction to the 1956 *Colaboradores e investigadores del Consejo Superior de Investigaciones Científicas*, the third of the CSIC's general publications, which took stock of the period between July 1945 (when the position of 'scientific collaborator' was created); May 1947 (when the creation of the position of 'scientific researcher' marked the 'first steps towards creating a scientific hierarchy'); and July 1951 (when the decree on scientific collaborators and researchers structured and stipulated 'the provisions concerning collaborators and researchers' in a single legal ordinance). This work, then, comprises an index, by CSIC board, of the scientific researchers and collaborators up to 1955.

The cover of *Colaboradores e investigadores* is understated, with the only illustration being the silhouette of the CSIC *Arbor Scientiae*, the metaphor for the productive union between the sciences of the spirit and the sciences of nature. This union is also represented in the book by the inclusion of the only two photographs that appear on its pages: the first shows the laboratory space, filled with scientific instruments being handled by two men in white lab coats, while the second features a man in a suit, sitting and reading in the library. This second space is represented by dozens of bound volumes displayed in an orderly fashion on the bookshelves, in front of which stands a woman. Located in the background, only her clothing is visible – a shirt, long skirt and shoes – but her face cannot be seen and there is no indication of the work she is doing (the faces of the male representatives, who are the focus of both photographs, are visible and they are (re-)presented as conducting research). 'Actors', 'objects' and 'places' in two images with four people: three men and one woman.

'A reality of life at the Spanish National Research Council' are the prescient words of the printed title that accompanies the two photographs.[2] Once again, a close reading or interpretation of the images provides some idea of the 'collective sensibility' of an era (Burke, 2001: 38–40), without falling into the trap of thinking that this reality was so homogenous as to be expressed in all its complexity in a single image. However, in practice, the ratio between the male and female scientific researchers and collaborators at the CSIC was not that far off.

Table I.1. Number of scientific researchers and collaborators by board (1945–55)

Board	No. of women	No. of men	Total
Raimundo Lulio Board for Theology, Philosophy, Law and Economics	0	8	8
Marcelino Menéndez y Pelayo Board for History, Philology and Art	1	9	10
Santiago Ramón y Cajal Board for Animal Biology and Medicine	2	26	28
Alonso de Herrera Board for Plant Biology and Agricultural Sciences	7	44	51
Alfonso el Sabio Board for Mathematics, Physics and Chemistry	4	28	32
Juan de la Cierva Board for Technical Research	2	18	20
Diego de Saavedra Fajardo Board for Geography and Bibliography	0	4	4

These numbers alone, which are taken from *Colaboradores e investigadores*, reflect the naturalised sexism inherent in the practices and discourses faced by women in the scientific field. However, these discourses and practices always maintained a distinction. For example, although obtaining a job did not, in practice, automatically entail a recognition of authority, in keeping with the spirit of the decree of 1951 on collaborators and researchers, the pages of this book do confer authority, if only according to 'the volume and importance of the published research and because of the diligent commitment'.[3] This scrupulousness, in turn, is also reflected in the brief biographies included for each researcher. These men and women, therefore, are given form on paper by their names and merits alone, translated here into publications and research stays abroad with which, in the words of the book, 'the reader will be able to judge the silent, fruitful labour of our research'.[4] With this in mind, Tables 2 to 6 list the women found working for the various CSIC boards.

Tables I.2–I.6. Female scientific collaborators (SC) and scientific researchers (SR), CSIC (1945–55)

Table I.2. Marcelino Menéndez y Pelayo Board for History, Philology and Art

Name	Position	Research stays	Total publications	Spanish publications	Foreign publications
Casado Lobato, María Concepción (1920–?)	SC 1954	Uni. Coimbra (1949) Uni. Paris (1950–1) Uni. Cat. Milan (1952) Uni. Amsterdam (1954) Uni. Bonn / Uni. Strasbourg (1954–5)	1	1	

Table I.3. Santiago Ramón y Cajal Board for Animal Biology and Medicine

Name	Position	Research stays	Total publications	Spanish publications	Foreign publications
Borrell Ruiz, Sara (1917–99)	SC 1949	Scotland (1946) England (1951) United States (1953)	15	14	1
Stamm Menéndez, María Dolores (1924–88)	SC 1951		11	11	

Table I.4. Alonso de Herrera Board for Plant Biology and Agricultural Sciences

Name	Position	Research stays	Total publications	Spanish publications	Foreign publications
Angulo Carpio, María Dolores (1915–?)	SR 1949	Est. Agro. Nac., Sacavém (Portugal) (1946) Botanic Institute, Cambridge (England) (1951) Max-Planck-Institut, Voldagsen (Germany) (1953)	14	14	
Aznar Ortiz, María del Pilar (1914–?)	SC 1946		2	2	
García Sanz, Dominica María Pilar (1920–?)	SC 1955		1	1	

Gónzalez Peña, Julia María (1925–?)	SC 1954	Scotland (1946) England (1951) United States (1953)	5	5	
Medina Ortega, Antonia María (1923–?)	SC 1951		3	3	
Pérez Mateos, Josefa (1904–94)	SR 1949	Landbouwhogeschool, Wageningen (Holland) (1948) Sediment-Petrographisches Institut, Göttingen (Germany) (1952)	22	21	1
Sánchez Calvo, María del Carmen (1919–?)	SC 1954		9	9	

Table I.5. Alfonso el Sabio Board for Mathematics, Physics and Chemistry

Name	**Position**	**Research stays**	**Total publications**	**Spanish publications**	**Foreign publications**
Cubero Robles, Mercedes (1924–69)	SC 1951		12	9	3
Domínguez Astudillo, María (1895–1985)	SC 1949	Study certificates, Universities of London, Cambridge and Durham (England)	16	16	
Martín Tordesillas, Isabel (1920–?)	SC 1955		6	6	

Molera Mayo, María Josefa (1921–2011)	SC 1949	Ramsay Fellowship Trust, Oxford University (England)	5	4	1

Table I.6. Juan de la Cierva Board for Technical Research

Name	Position	Research stays	Total publications	Spanish publications	Foreign publications
García Riquelme, Olga (1923–2012)	SC 1951	Lunds Universitet (Sweden)	5	4	1
Vigón Sánchez, María Teresa (1913–?)	SR 1949	Eidgenössische Polytechnische Schule (Zürich)	13	11	2

Officially, this is the first group of women who formed part of the CSIC research staff. For the most part, they were born in the 1920s, had excellent academic records and enjoyed the possibility of studying abroad. All of them embodied a suitable habitus, with a decisive social and cultural capital within a scientific field dominated by men. One such case is the doctor in Pharmacy, Sara Borrell Ruiz (1917–99), from the Institute of Bromatological Research, where she worked until September 1955, at which time she joined the CSIC Institute for Experimental Endocrinology directed by Dr Gregorio Marañón, with whom she had been collaborating since 1950.

Borrell Ruiz, a relative of the composer Isaac Albéniz (1860–1909), grew up in a well-to-do family and in a cultural and educational environment that fostered the study of science, music and art, and where academic excellence was demanded as the norm (Santesmases, 2017). María Teresa Vigón Sánchez (1913–?), the director of the photography and photochemistry section of the Daza de Valdés Optics Institute, represented cultural legitimacy and social stratification and was even a member of the sociological elite of the Franco regime. She was the daughter of General Juan Vigón, a minister in the Franco administration between 1940 and 1945 and president of the Board of Nuclear Energy (JEN) between 1951 and 1955, and sister of María Aranzazu Vigón, also a scientist, who was appointed director of the Instrumentation Physics Section of the JEN in 1954 under her father (Romero de Pablos, 2017).

On this list of CSIC women, María Teresa Vigón was one of the veterans and one of the only three female scientific researchers at that time, all of whom worked in the applied sciences. This speaks to both the importance of specific areas of science during the era of the self-sufficiency policies, as well as 'a bureaucratic sense, with value based only on official examinations or seniority',[5] which the CSIC wished to eliminate at the time, but was still very powerful in the mid-1950s. The two other female scientific researchers were the doctor of Natural Sciences and Pharmacy, Josefa Pérez Mateos (1904–94), a member of the Spanish Royal Society of Natural History since 1934 (Magallón Portolés, 2004: 333) and recipient of the Outstanding Doctorate Award from the Universidad de Madrid in 1944 (De Felipe Antón, 2002a: 486); and the doctor in pharmacy, María Dolores Angulo Carpio (1915–?), a member of the José Celestino Mutis Institute for Pharmacognosy and author of the book on cytogenetics, *Notas bibliográficas sobre agentes físicos en citogenética* (1955). In June 1954, before the text was published, she requested the manuscript from the CSIC in order to make, she said, 'a small modification to the title'.[6]

Significantly, at this time, publications served as the main mechanism to measure scientific productivity (González Blasco, 1980: 235–95). In this respect, these women – along with the member of the Teresian Institute and director of the Radiobiology Laboratory, María Domínguez Astudillo (1895–1985) – had the highest number of published works. According to the data in *Colaboradores e investigadores*, these scientific studies were largely published under the endogamic system of CSIC journals and monographs that so heavily contributed to the process of constructing scientific authority in Franco's Spain (García Naharro,

2022; Malet, 2008). In most cases, these women co-authored their works with men, figures of authority in the field, whose relationships often went beyond the text. Indeed, in practice, these women benefited from an advantageous social capital in an environment and at a time dominated by strong sexist prejudices. As just one example, the doctor of chemical sciences in the physical chemistry section in Seville, Mercedes Cubero Robles (1924–69), was married to Agustín Vioque Pizarro (1923–98), being herself a member of the Institute for Fats and disciple (and collaborator) of Luis Bru Villaseca. Bru Villaseca, in turn, was a disciple of Julio Palacios (Márquez Delgado, 2004: 117), and Cubero Robles co-authored nine of the twelve selected works on her curriculum vitae with him, including two published in 1952 in the *Journal of Chemical Physics* and *Proceedings of the Physical Society*.[7]

Like Luis Bru Villaseca, Miguel Antonio Catalán Sañudo was also suspended from his position after the Spanish Civil War. In his case, Catalán recovered his professorship at the Universidad de Madrid in early 1946, although he was not really allowed to return to what was deemed 'official' research until José María Otero Navascués appointed him head of the Spectrum Department at the Daza de Valdés Optics Institute in 1950 (Sánchez Ron, 1999: 319–27 and 1994: 360 and 345–94). There, he and his team demonstrated with their publications that there was some movement on the stagnant Spanish scientific-technical scene (García Naharro, 2020b). One notable member of that team at the Optics Institute was the doctor of chemistry, Olga García Riquelme (1920–2012), who, according to *Colaboradores e investigadores,* also co-authored four articles, two with her colleague Rafael Velasco and two with Miguel Catalán himself. Her case is similar to that of the doctor of chemistry at the Institute for Edaphology and Plant Physiology, Julia María González Peña, a pioneer in the study of electron microscopy in Spain (Rincón, 2010: 46). This scientist co-authored works with distinguished scientists of the time, such as José María Albareda, director of her institute and secretary-general of the CSIC, and Vicente Aleixandre Ferrandis, head of the physical chemistry section, also at the Institute for Edaphology. Her colleague, the chemist Antonia María Medina Ortega, recipient of the

Outstanding Undergraduate Award and Outstanding Doctorate Award, who was appointed head of the plant biochemistry section of the Institute for Edaphology in 1958, published alongside inorganic chemistry Professor Enrique Gutiérrez Ríos at the Universidad de Granada, for whom she worked as a teaching assistant.[8] Finally, the doctor in chemical sciences, María Josefa Molera Mayo, the daughter of a commander of the Carabinieri Corps who fought for the Nationalist faction and died in combat in 1937 and wife of Joaquín Hernáez Marín, a fellow researcher named head of the physical metallurgy section of the Department of Nonferrous Metals in 1960 (Carrascosa Santiago, 2012a: 384 and Crespo García *et al.*, 2009: 203), published in collaboration with Antonio Rius Miró, director of the Antonio de Gregorio Rocasolano Institute for Physical Chemistry.

These cases confirm some of the premises regarding women scientists under Franco: they had close ties to renowned male figures in Spanish science and/or were themselves members of the sociological elite of the Franco period; they had ideological legitimacy, having been neither purged nor exiled; and they were equipped with the cultural and social capital that was essential in a scientific field dominated by men. In short, they were brilliant professionals with ties to the CSIC, who – like many of their male colleagues – took advantage of the opportunity to do research and study abroad (González Blasco and Jiménez Blanco, 1979: 155–60) and, thus, enjoy the legitimacy associated with the international scientific community.

Chapter 7

Women Travelling: Research Residencies Abroad

After 1940, the Spanish National Research Council began to rely – in the words of the CSIC itself – 'indispensably' on 'collaborator "fellows"', along with the council's scientific collaborators and researchers. These grants, 'fully grafted onto the Council tree', were originally 'domestic fellowships to provide further education in Spain and in collaboration with the work of the Spanish National Research Council' and, once satisfactorily performed, raised the 'fellow' to the status of 'University and Secondary School Assistant (with the Right to Open Competition for Assistant Posts'. However, eligibility for a fellowship required 'having completed studies at University, the Technical School or a School or higher Centre of culture or technical specialization' and 'having completed a bibliographical or doctrinal research project with a professor who can guarantee the student's aptitude and abilities', in addition to 'being competent in at least two modern languages'.[1] At that time, these requirements clearly restricted access to a handful of students chosen from among the select circles of so-called high culture.

However, it was not until March 1946 that the Board of Cultural Relations (*Junta de Relaciones Culturales*, JRC), under the authority of the Ministry of Foreign Affairs, issued the first calls to grant 'fellowships to Chair Professors and Associate Professors to do research and study abroad'. University and secondary education chair professors, associate professors at higher level technical schools and collaborators within the Spanish National Research Council were invited to present their applications to 'the President of the Spanish National Research Council', since the CSIC institutes – or 'the Juan de la Cierva Board with regard to technical studies' – were responsible for providing reports that ultimately validated the candidacies that would be granted fellowships by the Board of Cultural Relations.[2] Once again, the dictatorship delegated part of the process of shaping scientific recognition to the CSIC, as the highest scientific institution in the regime, and according to the inner logic of the authoritarian 'social contract for science' in Franco's Spain: the council was tasked with deciding who was actually suitable to receive this funding. In the end, of the first forty-eight candidates for 'fellowships for Chair Professors and Associate Professors', not one was a woman.[3]

Also in 1946, a general call for 'fellowships for advanced studies abroad' and 'fellowships for research and professional technical development abroad' was issued, which again involved the Spanish National Research Council in the process 'before the Board of Cultural Relations takes its final decision'.[4] Originally offered for a length of six months to one year, the posts were later limited to three to six months. In June of that year, the candidates selected to move on to the language test included two women: the doctor in Chemical Sciences, María Domínguez Astudillo,[5] and the doctor in Medicine, Trinidad Peraza y Rodrigo de Villabriga,[6] both of whom applied for four-month research residencies in England, after being recommended by the Faculties of Science and Medicine in Madrid, respectively.

In short, of the forty-seven applicants for the research and professional technical development fellowships abroad, only two were granted to women, while of the seventy-two candidates for general fellowships for further studies abroad, only four went to women: Jerónima Moragues González, who received the good news from the dean of the Faculty of Sciences in Barcelona; Sara Borrell Ruiz, who was informed by the dean of the Faculty of Pharmacy in Madrid; Manuela López Díaz, informed by a chair professor in the Faculty of Pharmacy in Madrid; and Eva Martínez Martínez, who received her positive response from the dean of the Faculty of Pharmacy in Madrid. The first three women went to England, and the fourth to Belgium.[7]

A more in-depth understanding of this situation of gender inequality within the Spanish scientific field, along with the international scientific entanglements of an isolated Spain, can be gleaned from Issue 4 of the CSIC General Publications: *Becarios del Consejo Superior de Investigaciones Científicas* (1957). This volume covers the fellowships from the earliest years of the CSIC during the time of self-sufficiency in Spain to 1956, when the scarcity of the immediate post-war years began to recede and the regime made modifications to its international relations, ushering in a gradual and strategic process of opening up. The data from these decisive years regarding the various boards dedicated to the pure and applied sciences at the CSIC reveal a great deal about the profiles of the women of science who travelled outside Franco's Spain. The four boards of particular interest (Santiago Ramón y Cajal Board for Medical Sciences; Alonso de Herrera Board for Natural and Agricultural Sciences; Alfonso el Sabio Board for Mathematics and Physical Chemical Sciences; Juan de la Cierva Board for Technical Research) and their female fellows are examined individually below.

Table II.1. Santiago Ramón y Cajal Board for Medical Sciences

Name	
Sara Borrell Ruiz	1951 and 1952; six months and three months, respectively; chemical-analytical techniques, University of Cambridge, England
María Dolores Stamm Menéndez (1924–88)	1954 and 1955; four months each; Universität Tübingen and Max-Planck-Institut für Biologie Tübingen, Germany
María Luisa Alonso Puerta	1954; two years; virology, Institut Pasteur Paris, France
María Dolores Jurado Grau and **María Casal Wismer** (1929–2023)	June 1955; two months each; new techniques of Professor Fleisch (1933–2007); Physiology Laboratory, Université de Lausanne, Switzerland
Gabriella Morreale de Castro (1930–2017)	1955; one-year 'exchange fellowship'; the use of radioactive isotopes in human biochemistry; Leiden, Netherlands

Between 1940 and 1956, the Santiago Ramón y Cajal Board for Medical Sciences gave fellowships to 157 researchers, of whom only six were women.[8] One was María Casal, whose family had come from Switzerland, but who was born in Guillena, a town near Seville. Although raised as a Protestant, she ended up embracing Catholicism as a member of Opus Dei, for which she continued to work even though she moved to Zürich in the mid-1960s (Casal, 2019). The daughter of the Swiss electrical engineer Emilio Casal (an employee of Compañía Sevillana de Electricidad and Swiss consul in Seville) and educated in German and French schools, Casal came into contact with Opus Dei circles during her medical studies in Seville. In late 1952, this multilingual woman moved to Barcelona upon the express request of the Opus Dei acolyte and professor at the Barcelona Faculty of Medicine, Juan Jiménez Vargas (1913–97), to work in his department and translate articles on Physiology from German to Spanish, and Nursing from English to Spanish. In 1954, Jiménez Vargas was appointed dean of the Faculty of Medicine at the Universidad de Navarra, and twenty-five-year-old Casal accompanied him, to become the director of the new School of Nursing at the university, where she rendered 'all the spirit of service and sense of the human and Christian value of professional work that she had . . . learned from the devout Josemaría Escrivá' (Arribas and Serrano, 2001: 723–34).

When Jiménez Vargas left the deanship at the Faculty of Medicine, he was replaced by fellow Opus Dei member Eduardo Ortíz de Landázuri (1910–85), a man with close ties to another female researcher noted above, Gabriela Morreale de Castro, a woman whose scientific and academic vocation came from her family. The daughter of Emilia de Castro (1891–1975) and the Italian consul in Malaga, Eugenio Morreale (1891–1984), both of whom had degrees in the natural sciences, Morreale de Castro was educated in the German school in Malaga before beginning her studies in Chemistry at the Universidad de Granada in 1947, which she completed with the highest qualifications in 1951. That same year, she received her first CSIC fellowship to work in the

Physiopathology of Nutrition Laboratory, directed by Professor Eduardo Ortiz de Landázuri at the Granada Faculty of Medicine.

A practising Catholic who became a Spanish national in 1953 after marrying the CSIC scientist Francisco Escobar del Rey (1923–2015), Gabriela Morreale de Castro joined her husband for a research residency at the Universiteit Leiden academic hospital in the Netherlands, working with the endocrinologist Andries Querido (1912–2001). This residency, which was arranged through José María Albareda, continued for three years, during which time Morreale de Castro was appointed *hoofd-assistant* (laboratory director) during the 1957–8 academic year. It was also at this time that she co-wrote her first international publications in collaboration with her husband, yet another example of how some scientific couples cooperated in a way that benefitted them both. On her return from the Netherlands, and once again with the assistance of Albareda, Morreale de Castro joined the Biological Research Centre (*Centro de Investigaciones Biológicas*, CIB) in 1958, having reached the level of CSIC collaborator one year earlier. In 1962, she became a CSIC scientific researcher, and between 1963 and 1975, she headed the Thyroid Study Section of the Gregorio Marañón Institute at the CSIC. In 1970, she rose to the rank of research professor, the highest scientific position in the council (Carrascosa Santiago, 2012a: 490–2; Moreno *et al.*, 2019).

Table II.2. Alonso de Herrera Board for Natural and Agricultural Sciences

Name	
Mercedes Aguado Marín (?–1999)	1952; one and two months respectively; improvement of plant facilities; Estação Agronómica Nacional, Sacavém, Portugal and medicinal plant improvement facilities; Rijks Landbouw Hoogeschool Wageningen, Netherlands
Concepción González Moreno	1946; one year; plant physiology; University of California Berkeley, United States

María Teresa Rodríguez Mellado	1947; one month and ten days; Serviços Geológicos de Portugal, Portugal
María Dolores Angulo Carpio	1946–53; one year and two months; genetics; Instituto de Alta Cultura Sacavém with Professor Cámara, Portugal; one and a half months; Estação Agronómica Nacional, Sacavém, Portugal; four months; Botany School, University of Cambridge, England; three months; Max-Planck-Institut Voldagsen, Germany
Josefa Pérez Mateos	1948 and 1952; three months with Professor Edelman, Rijks Landbouw Hoogeschool Wageningen, Netherlands; one three-month exchange fellowship in Göttingen, Germany
María del Carmen Sánchez Calvo	1952; five months; Göttingen, Germany
Julia Lérida García	1950; one year; Estação Agronómica Nacional, Sacavém, Portugal
María Aparicio Gallego	1950; one month; edaphology; Geneva, Switzerland
María Luisa Bermejo Rosciano	1950; one-and-a-half months; Estação Agronómica Nacional, Sacavém, Portugal
Josefina Bru Villaseca	1952; one month; library organisation; Estação Agronómica Nacional, Sacavém, Portugal
María Blanca Ruiz del Castillo Navascues	1954; one month; Estação Agronómica Nacional, Sacavém, Portugal
María Luisa García García	1956; three months; Estação Agronómica Nacional, Sacavém, Portugal
Elena Paunero Ruiz (1906–2009)	1949; twenty days; herbs and libraries; Rabat and Algiers; 1952; two months; London
Isabel García Acha	1952; one-and-a-half months; plant analysis; Universiteit Leuven, Belgium; 1954; three months, England

Julia María González Peña	1952; three months, respectively; Ghent, Belgium; Delft, Netherlands
Elena Humbert Gamhon	1951; forty days; Austria and Switzerland; 1952; three months; soil fauna; Graz, Austria; 1955; one month; visiting the laboratories of Professor Kühnelt; Vienna, Austria
Concepción Bonet Muñoz (1916–2014)	1953; four-month exchange fellowship; coal megaspores; Geologisch Bureau Heerlen, Netherlands
Elisa Ibarrola Muñóz	1951 and 1954; ten and eight months, respectively; Eidgenössische Polytechnische Schule Zürich, Switzerland
Narcisa Martín Retortillo (1910–2003)	1951; three months; Laboratorium für Erdbereitung Hamburg, Germany
María Carmen González Vázquez	1953; one year and six months; Laboratoire de Cryptogamie, Paris, France
María Paz Rodríguez Álvarez	1954; one year; Lunds Universitet, Sweden
Genoveva Tejerina Domínguez	1953; two months; Groningen, Netherlands
Antonia María Medina Ortega	1952; three months; Universiteit Leuven, Belgium; three months, Rijks Landbouw Hoogeschool Wageningen, Netherlands; 1955; one year and one month; England
Ana María Sánchez de Rivera Alfaro	1951; four months; bibliographical studies with Professor Angulo; England
María del Carmen Sánchez Conde	1955; one year; Universität Bonn, Germany (see Table II.4)
Concepción Menéndez Domingo	1955; nine months; phycology studies with Professor Feldmann; France
María Pilar Garrido García	1956; six months; Faculty of Pharmacy; Universidade do Porto, Portugal

Josefa Menéndez Amor (1926–85)	1954; three months; 1956; two months and fifteen days, respectively with Professor Sitter and Professor Florschütz; Netherlands and the Cantal department, France

From 1940 to 1956, the Alonso de Herrera Board for Natural and Agricultural Sciences awarded fellowships to 175 researchers, of whom twenty-eight were women.[9] Once again, this group was made up of women raised in settings that fostered science, like Elisa Ibarrola Muñoz – the daughter of the doctor of Medicine and Surgery and director of the National Psychotechnical Institute, Ricardo Ibarrola Monasterio (1903–82) – who attended secondary school at the Lycée Français de Madrid (Quintana Fernández, 2010), before becoming a geologist and eventually the director of the CSIC's Institute for Economic Geology. Furthermore, some scientific women had strong religious beliefs, such as Mercedes Aguado Marín, who obtained a doctorate in agronomic engineering and the exact sciences as a devout believer, ordained as a Slave of the Sacred Heart of Jesus, and Josefa Menéndez Amor, who, as a fellow at the José de Acosta Institute, contributed to the monograph on evolution published by the journal *Arbor* in 1951, in which she acknowledged the intervention of God in her article on evolution in the plant kingdom.[10]

However, this group also included some women whose careers were connected to the now defunct Board for Advanced Studies and Scientific Research. Such cases include Narcisa Martín Retortillo, a pioneer in the field of soil humic substances who had formed part of the National Institute of Physics and Chemistry during the Second Republic (De Felipe Antón, 2002a: 485) (Alcalá Cortijo and Magallón Portolés, 2008: 161), and Elena Paunero Ruiz, a member of the JAE mycology group who was made assistant in the Herbarium Section of the botanic garden in 1946 after the CSIC was founded, and appointed head of the Agrostology Laboratory in 1948 (Valdés, 2009).

This large and diverse group of women all shared the privilege of travelling to study abroad, a rite of passage for the women who had acquired the habits of independent, free individuals, with the subsequent access to academic prestige and professionalisation (Santesmases Navarro de Palencia, Cabré i Pairet and Ortíz Gómez, 2017: 397). Some of these women, like Josefa Pérez Mateos, began to travel as early as the second half of the 1940s; in 1948, for instance, she spent three months working in the Netherlands with the scientist Cornelis Hendrik Edelman (1903–64), a specialist in soil sciences. A pioneer in the field of sedimentary petrology, Pérez Mateos directed two sections and was a member of the governing board of the CSIC Institute for Edaphology, where she began as a collaborator, before moving up to scientific researcher (1949–70) and research professor until she retired in 1975 (De Felipe Antón, 2002a: 485–6).

Even earlier, in 1946, María Dolores Angulo Carpio travelled to Portugal to work with António Pereira de Sousa da Câmara (1901–71), the founder and first director of the Estação Agronómica Nacional (Saraiva, 2016: 43–71) and a renowned figure in the field of cytogenetics (De Felipe Antón, 2002a: 486). Thanks to his friendship with Albareda and their mutual interest in Agricultural Chemistry and applying genetic improvements to the agrarian economy (González González, 2002: 115–16), Sousa da Câmara was included as an honorary board member of the CSIC,[11] which strengthened the ties between the two institutions. In 1950, Angulo returned to Portugal, where she hosted María Luisa Bermejo Rosciano, a postgraduate research assistant at the CSIC until May 1954.[12] María Luisa García García, a research assistant who stepped down from her post in September 1959, was another fellowship recipient who passed through Sacavém, as did Josefina Bru (whose brother Luis Bru Villaseca, a reinstated disciple of Julio Palacios, was a chair professor of Theoretical Physics at the Universidad de Sevilla beginning in 1942, before moving to the Universidad de Madrid in 1955), Julia Lérida García and María de la Blanca Ruiz del Castillo Navascues.

In 1955, Concepción Menéndez Domingo received a fellowship to extend her studies in Phycology in Paris with the French phycologist Jean Feldmann (1905–78), while Josefa Menéndez Amor – head of the micropalaeontology section of the Lucas Mallada Institute, who finally reached the rank of scientific researcher at the CSIC[13] – followed suit by going to work in the Netherlands with the bryologist Peter Arnold Florschütz (1923–76). She was preceded by the geologist Concepción Bonet Muñoz, who travelled to the Netherlands in 1953 to work under the direction of geologist S. J. Dijkstra on carboniferous megaspores, the subject of her doctoral thesis in Spain, which was awarded the highest qualifications in 1955. Her dissertation was directed by Bermudo Mélendez Meléndez (1912–99), a chair professor in palaeontology and geology at the Universidad de Madrid and head of the palaeontology section of the Lucas Mallada Institute. In 1941, the regime had awarded Meléndez a medal in recognition of the time he was held captive in the Barcelona Model Jail during the Spanish Civil War. Bonet Muñoz worked with Meléndez throughout the 1940s, teaching classes in the natural sciences in his private academy. However, Meléndez was best known for his staunch defence of the finalist, theistic evolutionism tolerated by the Franco regime and championed by the Arbor Group within the CSIC (Florensa, 2017; Otero Carvajal and López Sánchez, 2012: 259–316).

While Concepción Bonet worked in the shadows of Bermudo Meléndez during the 1940s, María Teresa Rodríguez Mellado was under the protection of Francisco Hernández Pacheco (1899–1976), a chair professor at the Universidad de Madrid with whom she published her first work in 1947 and who, in the opinion of botanist Luis Crespí, operated at the time 'surrounded by a court, an estate of sorts' (Otero Carvajal and López Sánchez, 2012: 309). Isabel García Acha, in turn, enjoyed the support of her future husband, Julio Rodríguez Villanueva (1928–2017), and both were members of the Edaphos Club, under the protection of José María Albareda (Rodríguez Villanueva and García Acha, 2002). Again, proximity to the male scientists and scientific figures at the CSIC gave the female scientists crucial social capital in a setting dominated by men and strong sexist preconceptions. Like Concepción Bonet, all these women had to circumvent biases, discrimination and barriers in order to pursue their own research lines.

Beginning in the second half of the 1950s, Bonet specialised in photogeology, working both at the Universidad de Madrid and on the Nuclear Energy Board (JEN) in 1957. In the mid-1960s, she joined the Materials Division of the Directorate-General for Roads, a section of the Department of Public Works in Madrid, thus becoming a pioneering woman in the world of professional Geology in Spain.[14]

Table II.3 Alfonso el Sabio Board for Mathematics, Physics and Chemistry

Name	Research stays
Amelia D'Ocón Asensi (?–2011)	1946; one month; organic chemistry research; Switzerland and England
María Domínguez Astudillo	1949; five months; University of Cambridge Radiochemical Laboratory, England
Margarita Bernis Madrazo	1951; three weeks; electronics professional development; London, England
María Josefa Molera Mayo	1950; one year; Physical Chemistry Institute; University of Oxford, England
Antonia Martín-Tesorero Álvarez	1951; calculation of optical systems; Boston, United States
Mercedes Cubero Robles	1952; six months; determination of molecular structures by X-ray diffraction; Universität Graz, Austria
Baldomera Velázquez Sánchez	1955; one year and seven months; Institut Français du Caoutchouc, France
María Luisa Canut Ruiz	1956; six months; X-ray diffraction studies at low and high temperatures; University College London, England

From 1940 to 1956, the Alfonso el Sabio Board for Mathematics, Physics and Chemistry awarded fellowships to 128 researchers, of whom only eight were women.[15] One was Amelia D'Ocón Asensi, a PhD student and the second wife of Manuel Lora-Tamayo, a woman educated at the select Instituto-Escuela and the Residence for Young Ladies (*Residencia de Señoritas*) in Madrid, who joined the Royal Spanish Natural History Society in the mid-1920s. The daughter of the industrial engineer Emilio D'Ocón Cortés (1891–1972), she had worked as a primary education inspector before the Civil War (Magallón Portolés, 2004: 308) like her mother before her, Amelia Asensi Breviá (1893–1974), a school inspector in Toledo who also enjoyed several research residencies abroad funded by the JAE (Marín Eced, 1991: 50–2). In July 1945, Amelia D'Ocón sat for the open competition to fill a teaching position at the School of Industrial Technicians,[16] and in 1946 she embarked on a year abroad that turned out to be a very special time for her, as she both married and defended her doctoral thesis – directed by Lora-Tamayo – on the aromatisation of propenylbenzene and maleic anhydride adducts.

At the end of the decade, in 1949, María Domínguez Astudillo – better known at the Antonio de Gregorio Rocasolano Institute for Physical Chemistry as 'Miss Astudillo' (Rico, 2009: 186) – went abroad, in this case to the University of Cambridge in England to further her studies at the Radiochemical Laboratory of Alfred Gavin Maddock (1917–2009). Two years later, Margarita Bernis Madrazo, a woman who graduated from the Instituto-Escuela in 1932 and obtained a degree in the sciences from the Universidad Central de Madrid in 1936, followed a research residency in the United States, thanks to a fellowship at Bryn Mawr College (academic year 1949–50)[17] with a trip to England in 1951, where she continued her study of electronics, as did María Josefa Molera Mayo, who studied in the Physical Chemistry Laboratory at the University of Oxford in England under the direction of the future Nobel Prize winner in Chemistry, Cyril Norman Hinshelwood (1897–1967). According to the sociopolitical logic of the Cold War, the English-speaking world was the new focus of attention, as English became consolidated as the *lingua franca* of science in light of the political and linguistic collapse of German, with Russian rapidly moving into second place in the mid-1960s (Gordin, 2015).

While Molera travelled as a single woman, Mercedes Cubero Robles was already married – to Agustín Vioque Pizarro, an important member of the CSIC Institute for Fats – as was Antonia Martín-Tesorero Álvarez, who travelled abroad as a newlywed, having married the chair professor of Industrial Physics at the Universidad de Madrid, José García Santesmases (1907–89) in December 1950. The couple spent 1951 in Boston, where she studied the calculation of optical systems and he worked in the Harvard Computation Laboratory under the direction of Howard H. Aiken (1900–73). In 1956, María Luisa Canut Ruiz spent six months at University College London in England, while her husband, José Luis Amorós Portolés (1920–2001), a researcher at the CSIC until 1954, prepared to become chair professor of Crystallography at the Universidad de Madrid, where he would stay until 1966 (López-Acevedo and López Andrés, 2001), the same year that Canut, then a scientific researcher at the council, stepped down from her position at the CSIC.[18]

Table II.4. Juan de la Cierva Board for Technical Research

María Teresa Vigón Sánchez	1947; ten months; with Professor Eggert; Eidgenössische Polytechnische Schule Zürich, Switzerland
Piedad de la Cierva Viudez (1913–2007)	1948; optical glass with Professor Hubbard, Professor Halmer and Professor Heind; National Bureau of Standards; Washington DC, United States
Olga García Riquelme	1952; eighteen months; excitation of doubly ionised spectra with Professor Edlen; Lunds Universitet, Sweden
Victoria Redondo Mena	1954; fifteen months with Professor Slevogt; Technische Universität Berlin Optisches Institut, Germany

María del Carmen Sánchez Conde	1955; with Professor Neuhaus and Professor Konopicky; Institut für Mineralogie, Forschungsinstitut der Feuerfest-Industrie, Universität Bonn, Germany
María Teresa Sardiña Gallego	1956; four months with Professor Mathieu; Université Paris-Sorbonne, France
Juana Bellanato Fontecha	1956; nine months; with Professor Mecke; Institut für Physikalische Chemie, Universität Freiburg, Germany (Bellanato was granted the fellowship despite the misgivings of a colleague)

From 1940 to 1956, the Juan de la Cierva Board for Technical Research awarded fellowships to 103 researchers, of whom the seven women listed above were the only female recipients.[19] Again, well-positioned figures were behind the names of these female fellows, like María Teresa Vigón Sánchez and, of course, Piedad de la Cierva, who was born into an elite family from Murcia and related to both Juan de la Cierva Peñafiel (1864–1938) and Juan de la Cierva Codorníu (1895–1936), spoke French, English and some German and was the first woman to work in the CSIC Institute of Physics and Chemistry. The security of her position was due to both her commendable professional activity and her political ties, as well as the friendships between her family and the new leaders (Alva Rodríguez, 2016: 6). However, before her residency with the National Bureau of Standards in Washington DC in 1948, de la Cierva – who obtained her first degree in 1932 with the highest qualifications and earned a PhD in Chemistry in 1934 – received funding from the JAE in 1936 to work at the Universitetet Institut for Teoretisk Fysik in Copenhagen with Professor Georg Charles von Hevesy (1885–1966), who was awarded the Nobel Prize in Chemistry in 1943 (Magallón Portolés, 2004: 240–2; Romero de Pablos, 2017: 325–6). In short, Piedad de la Cierva was always well connected. Her first association with the Board for Advanced Studies and Scientific

Research was facilitated by a letter of introduction from Professor Antonio Ipiens Lacasa, chair professor of general chemistry at the Universidad de Valencia, to Julio Palacios. Palacios agreed to direct her thesis and from October 1932, when she arrived in Madrid, to 1936, Piedad de la Cierva worked in the X-ray Section of the National Institute of Physics and Chemistry.[20] During that same time, María del Carmen Sánchez Conde also worked in the sciences, specifically in the Chemical Analysis Laboratory at the Faculty of Pharmacy in Madrid under the direction of José Casares, who would run the Institute for Physics and Chemistry at the CSIC after the Spanish Civil War (see Table II.2) (Magallón Portolés, 2004: 215). The war, which changed the lives of so many Spaniards, also affected the professional future of de la Cierva: shortly after the Civil War broke out, she was expelled from the Rockefeller Building by a working committee and forced to take refuge – through the mediation of her uncle, Ricardo de la Cierva y Codorníu – in the Norwegian embassy in Madrid. There she met José María Otero Navascués, a key scientific figure during the Franco regime, who would play a decisive role in her professional life (Pimentel Igea, 2020: 312–41), as he did with María Teresa Vigón and Juana Bellanato.

> When I received the fellowship to further my studies in Germany, a colleague advised the director of the Optics Institute, at that time José María Otero Navascués, not to let me go, because, according to him, I was prone to making a tempest in a teapot. That same colleague confessed this to me a year later. Thank goodness that the director had faith in me.[21]

This is how Juana Bellanato Fontecha remembered the misgivings of a male colleague when she, a woman, received a fellowship in 1956 to work at the Physical-Chemistry Institute of the Universität Freiburg in Germany with Professor Reinhard Mecke (1895–1969). Bellanato, then a scientific collaborator at the Juan de la Cierva Board, had defended her doctoral thesis two years earlier in 1954, after first obtaining the sponsorship of the rehabilitated Professor Miguel Catalán. This began her long, successful career with the CSIC.

Apparently, for some time, there was a critical misunderstanding between Miguel Catalán and Piedad de la Cierva. Catalán believed that her father, Juan de la Cierva y López (1888–1966), had been involved in the purge against him to benefit his daughter. Piedad de la Cierva herself denied this, assuring Catalán that her father had really made arrangements to try to reinstate him as soon as possible for Spanish science (Pimentel Igea, 2020: 332), which finally happened when José María Otero Navascués appointed him head of the Spectrum Department at the CSIC in 1950 (Sánchez Ron, 1999: 319–27 and 1994: 360 and 345–94).

During this time, Olga García Riquelme had established herself as an important member of Miguel Catalán's team at the Daza de Valdés Optics Institute, and in January 1952, she went to Sweden to further her education with the acclaimed Professor Bengt Edlén (1906–93), a specialist in spectroscopy. Miguel Catalán was also working with Edlén, his colleague at Lunds Universitet, when the famous Spanish discoverer of multiplets – a key step in the development of quantum theory – suddenly died in Madrid on 11 November 1957 (Sánchez Ron, 1994: 392). One year earlier, in 1956, secondary school physics and chemistry professor María Teresa Sardiña Gallego studied molecular spectra in Paris at the Sorbonne for four months alongside Professor Mathieu. In March 1961, she was given an extended leave, 'without position or payment' at the Cervantes Secondary School in Madrid and began to provide her services to the Board of Nuclear Energy (JEN).[22] At that time, nuclear energy was a priority for the Franco authorities, and the JEN was the preserve of José María Otero Navascués (Presas i Puig, 2021), the already mentioned Catholic mariner who took shelter in the Norwegian embassy during the Spanish Civil War and now stood at the helm of nuclear development in Spain.

Clearly, life circumstances, education, class and gender can never be underestimated as variables that explain biographies in the scientific world throughout history. This was especially true during the Franco period, beneath the promulgated official image of an 'emphatically Catholic' science[23] that reinforced the concept of the realm of science as a privileged place for the male/subject/knower, but where some women also knew how to find a place for themselves. These women had brilliant careers and records, many having been educated in select institutes and raised in well-to-do Christian environments, where the demand for academic excellence was the norm. All of them were also well situated and enjoyed social capital that spread throughout networks of recommendations, social relationships and conjugal relationships with male scientists.

However, judging from the official data, during a period of sixteen years, fellowships were granted to forty-nine women[24] from the CSIC's boards of pure and applied sciences. This number is not negligible, considering that out of the 113 individual grants given to women by the JAE between 1908 and 1934, only eighteen (16 per cent) were set aside for scientific studies.[25] Overall, the CSIC numbers for women were low when compared to the 514 male researchers who obtained fellowships during the same period; women fellows from these boards represented only 8.7 per cent of the total recipients. These numbers correspond to both the dynamics of the scientific field and to the professional segregation – horizontal and vertical – that prevailed in the CSIC (Alcalá Cortijo, 1996: 61–76).

The data also provide other information of interest. Of these female fellows, more than half (57 per cent) came from institutes dedicated to the natural and agricultural sciences, centres that had a higher number of researchers in general, also research areas with more women scientists than others. Again, these characteristic dynamics of the scientific field were not limited to Spain but were also international; of the various fields that comprise so-called 'science', some contained a higher proportion of women than others. For example, in fields like engineering and the technical and applied branches – which are even more masculinised, perhaps because they are more strongly associated with industry and the army (Noble, 1999: 209–28; and 1992) – the women there were usually viewed as pioneers or exceptional cases. This could be due to the fact that access to this field was traditionally more complicated for women, and, in many cases, they were family members, wives or friends of male scientists who already worked in them. In the words of Margaret W. Rossiter (Rossiter, 1997: 171), these were 'peripheral fields' of science for women.

One final, important aspect of women's work within the sphere of official science in Franco's Spain concerns publishing. Although publications did not appear to play as decisive a role as seniority in the in-house promotion of researchers during the dictatorship (Santesmases Navarro de Palencia, 2000: 137–8), they were always one of the primary mechanisms for scholarly legitimation and scientific prestige during the Franco period as well. Understood as the legitimate vehicle of scientific knowledge – the official and institutional locus of scientific authority – the CSIC's scientific journals were an active tool behind the bounding, shaping and maintenance of the Spanish scientific community under Franco (Gracía Naharro, 2020b), while also serving as a vehicle for the 'outward revelation' of the purported 'inner vitality and dedication' of the leading Spanish scientific institution: the Spanish National Research Council.[26]

Accordingly, the following chapter analyses one of the CSIC's main periodicals, the *Revista de Ciencia Aplicada*, to better understand the female scientists who sought to make a name for themselves in these 'peripheral fields' of science for women. Considered the 'central journal' of the Juan de la Cierva Board, it serves, in Lora Tamayo's words, as a prime example of an objective scientific journal (Nieto-Galan, 2019: 170, note 29). Not only did it include research articles amongst its pages, but also the types of section that characterised this kind of printed product of science: a 'National Information' section, a 'Bibliography' section and a bibliographical index of the 'books and leaflets' received by the Juan de la Cierva Board, as well as a 'Foreign Information' section, which brought the main advances in scientific research in Europe, the United States and Japan to the reader, while buttressing the scientific organisational model and image championed by the regime (García Naharro, 2017b). This simply serves as another example of the revelatory nature of marginal data and how every available source is, again, susceptible to being approached from a different perspective.

Consequently, the following chapter focuses on some of the – apparently – less significant details of the scientific articles in the *Revista de Ciencia Aplicada*. Leaving its contents aside, the chapter looks at who published in them (gender, position, number of times they wrote) and the extent to which and in what conditions women were involved, trying to develop a preliminary profile of their names, which are, indeed, the protagonists of the investigation. This approach owes much to the fact that, in modern science, every author requires a real name that establishes them as the person who is responsible for the assertions made. Ultimately, however, these assertions acquire all their value, recognition and credibility through their material relationship with other names, be they institutions, groups or authors of other publications with which the article engages. These names are also located in different spaces within the text, and even printed using different typographies, as in the case of footnotes, acknowledgements and dedications. This entire economy and geography of names, whether arbitrarily or conscientiously placed in the printed space, endorse and shape the 'author function', and, in the end, vouch for the reliability of the contents in the eyes of the scientific community (Biagioli, 2006 and Latour, 1992: 21–61; Foucault, 1998; Wilson, 2004).

Chapter 8

Women Publishing: Articles by Female Authors in the Journal *Revista de Ciencia Aplicada*

Although the *Revista de Ciencia Aplicada* began to be published in 1947, the first articles by female researchers did not appear in its pages until the early 1950s. An article on liquid-liquid extraction ('La extracción líquido-líquido aplicada a sistemas de cuatro componentes', 1952), for instance, was published in issue 27 – without any institutional affiliation – under the name of the doctor in chemical sciences, Maria Alicia Crespí (1922–2012).[1] This pioneering woman, who was raised in an intellectual environment, first worked in the National Institute of Industry's Spanish Pyrite programme before moving to the Nuclear Energy Board (JEN) in 1957. In 1970, she was awarded a fellowship from the Juan March Foundation, and in 1975 became a chair professor of electrical engineering at the School of Architecture in Madrid, making her one of the first female chair professors at a high technical school in Spain.[2]

With her 1952 article, Crespí appeared as the sole author, which was not the norm for female authors who published in the journal. Usually, in such author bylines – which were always validated by the author's qualifications – the women's names were accompanied by those of co-authoring colleagues who were higher up in the hierarchy. For example, Fermina González Pellissó (1912–2017), who had a degree in chemistry and worked as a researcher at the Institute for Fats, co-authored her work on preparing olives ('Estudios sobre el aderezo de aceitunas verdes', 1953) with the head of the chemistry section, José María Rodríguez de la Borbolla, and researcher Carlos Gómez Herrera. Although the 1953 piece was the ninth such article published under this generic title in a variety of journals, it was the first that featured González Pellissó's name in this journal.[3] Around the same time, researcher Felisa Núñez Cubero (1924–2017) published an article on permanent magnets ('Imanes permanentes', 1954),[4] which was co-authored by her thesis director at the Universidad de Valladolid, Salvador Velayos, a disciple of the physicist Blas Cabrera and erstwhile researcher at the JAE's National Institute of Physics and Chemistry, who was working in the Magnetics Section of the Alonso de Santa Cruz Institute of Physics at the CSIC (Baltá Calleja, 2009: 176).

Two years later and working alone, Felisa Núñez Cubero published a study on demagnetising magnetic materials ('La curva de desimanación de materiales para imanes', 1956), in which she expressed her 'gratitude to Prof S. Varela for the great interest he showed in directing my work and to the Spanish National Research Council, thanks to whose help it was made possible'.[5] In Núñez's case, she did not only connect her work to the credibility of male scientists and institutions, but also turned to a linguistic recourse for emphasis: the use of the masculine gender to describe herself. For instance, in her articles Núñez appears as a '*profesor adjunto*' (male associate professor) and '*becario*' (male fellow), using the generic masculine instead of the feminine gender in Spanish, unlike some other female researchers cited in the journal. The masculine gender was also used as a norm by Josefa Molera, Aurora Sampedro Piñeiro and Carmen Sánchez Conde, not only concealing their status as women, but also reflecting a well-adapted habitus (Bourdieu and Wacquant, 1992: 115–40) and even an inclusion strategy (Santesmases, 2017: 294) in the logic of a scientific field that Felisa Núñez Cubero and her female colleagues knew very well.

In fact, Felisa Núñez received a fellowship from the Alfonso el Sabio Board for Mathematics, Physics and Chemistry in the late 1950s to spend a year with Professor Leslie Fleetwood Bates (1897–1978) at the University of Nottingham in England, returning there in 1961 with a Ramsay Fellowship to finish some studies that would end up being cited in Bates's book *Modern Magnetism* as well as *Magnetism*, a four-volume collection that comprised the authoritative encyclopaedia on magnetism edited by George T. Rado and Harry Suhl.[6] During that same time, her work on sintered material and magnets ('Material sinterizado para imanes', 1960) was also published in the *Revista de Ciencia Aplicada*, with Núñez appearing again as a '*profesor adjunto*/male associate professor' of physics at the Universidad de Madrid,[7] where she had received her doctorate two years earlier. In 1964, she became the first female chair professor at the newly created School of Technical Engineering in Madrid.[8]

Noblesse Oblige: The Order of Names

In 1956, María Ángeles Sicilia Lafont (1914–2016), a graduate in chemical sciences and fellow in the Industrial Fermentation Section of the Juan de la Cierva Board, co-wrote an article with a colleague, Ricardo García Porras, also a fellow, and the director of the section and doctor in chemical sciences at the Universidad de Madrid and University of Manchester, José María Garrido Márquez. Next to the title of this article on fat production ('Producción de grasas por microorganismos', 1956),[9] the names of the two fellows appeared first, followed by the name of director. This practice of ordering the names corresponded to an etiquette convention that placed the most eminent scientist – often the laboratory director – in the last position. Although this formality could theoretically give young researchers some degree of visibility, in practice the work ended up being referred to in academic circles by the name of the author with the highest reputation, what Robert K. Merton dubbed the 'Matthew effect'. Indeed, at times, the big names appeared even though these academics had not actively participated in the work, simply because their mere presence bequeathed symbolic capital on the text as a whole (Biagioli, 1998: 3–16; Garfield, 1982: 62). In this regard, Manuel Lora-Tamayo himself remarked:

> There is a lack of morality on the part of the researcher who, having confused being the generic director of an Institute or Section with being the specific head of some research project, coerces his collaborators to appear as a participant in all the studies, and nevertheless presents a high degree of ethics as the director of a specific study postpones his name in such publications, when his work initiative was surpassed by that of the collaborator. (Lora-Tamayo, 1953: 19)

José María Garrido Márquez, who served as deputy counsel to the Juan de la Cierva Board in the 1960s,[10] was one such ethical figure, a man with a powerful surname who regularly made use of this name placement etiquette convention by postponing his name in scientific article bylines. The son of a doctor and chair professor of Surgery, Fermín Garrido Quintana (1868–1936), one of the most notable royalists in Granada during the 1930s,[11] José María Garrido spent two years at the Manchester Technical School, thanks to a council fellowship granted him in October 1950.[12] While in Manchester, he worked with Thomas Kennedy Walker (?–1970), an authority in the field of fermentation processes, with whom he published a number of articles in international journals such as the *Journal of the Science of Food and Agriculture* in the second half of the 1950s (García Naharro, 2020b).

In 1956, the name José María Garrido appeared again in the *Revista de Ciencia Aplicada*, this time following the name of Concepción Llaguno (1925–2010), a graduate in chemical sciences and fellow in the Juan de la Cierva Board's Industrial Fermentations Section. The two appeared as co-authors of an article on fumaric acid ('Producción de ácido fumárico por fermentación', 1956),[13] which also happened to be the title of Concepción Llaguno's dissertation, written under the direction of José María Garrido and read at the Universidad de Madrid in 1959. Unlike most of the women who have appeared in this book, Llaguno's mother was a teacher who was subjected to political repression and her father a postal service civil servant. During the years of her education, she combined her studies in chemical sciences at the Universidad Central with her job with the postal service,[14] and in 1953, after having obtained her degree, she began to work in the industrial fermentations section, a member of the group that won the Juan de la Cierva Prize in 1961.[15]

José María Garrido was a prolific author. His name also appeared following those of women like Margarita Buendía, a doctor in chemical sciences and collaborator with the industrial fermentations section, in 1958[16] and 1959,[17] and Ilse Schnabel Bender (?–1999), a graduate in chemical sciences, with whom he collaborated in 1964 – the same year that she obtained her PhD – and in 1965.[18] Again in the mid-1960s, Garrido appeared as a co-author of another article, following three female researchers: María Josefa Fernández, his doctoral student María Dolores Cabezudo – both graduates in chemical sciences and fellows in the section[19] – and Concepción Llaguno, at this time a doctor in chemical sciences.[20]

Like Garrido, other illustrious male authors associated with the journal also often appeared among the bylines, in the final position after their researchers. Notable cases include the chair professor José Ignacio Fernández Alonso (1917–99),[21] head of the Valencia section of the Department of Silicates, and Eduardo Primo Yúfera (1918–2007), director of the Department of Plant Chemistry (Ayala-Gascón, Aleixandre-Benavent and Gandía-Balaguer 2012). In 1958, for example, R. García Álvarez, a female graduate in chemical sciences and fellow in the Juan de la Cierva Board Department of Plant Chemistry appeared before José Royo Iranzo and Eduardo Primo Yúfera as one of the authors of an article published in the *Revista de Ciencia Aplicada* on concentrated orange juice ('Conservación a diversas temperaturas del zumo de naranja concentrado', 1958).[22]

However, this particular gendered order of names was not always the norm at the journal. In fact, in the mid-1950s, Clara Díez de Bethencourt, a graduate in chemical sciences, fellow in the industrial fermentations section and pioneer in wine chemistry research in Spain (Martínez Rodríguez, Carrascosa Santiago and Santiago Blanco, 2016: 202–04), co-authored an article on wine ('Algunos componentes fundamentales de los vinos de Rioja', 1957), appearing after Ildefonso Mareca Cortés (1924–2021).[23] Similarly, Carmen Janer del Valle, a doctor in chemical sciences and collaborator at the Institute for Fats and their Derivatives, appeared after Carlos Gómez Herrera, the head of the chemistry section at the institute, above their co-authored paper on olive oil.[24] The assistant professor of analytical chemistry in the Faculty of Sciences María Luisa Rexach-Martínez De Lizarduy (1929–2019), who was also a Juan de la Cierva fellow, also appeared after the names of Fernando Burriel Martí and Juan Ramírez Muñóz, in the list of authors of an article on sulphate ions ('Interpretación de algunas anomalías en la determinación de iones sulfato por fotometría de llama indirecta', 1958).[25] Rexach-Martínez De Lizarduy, an outstanding student at the La Milagrosa school run by the Franciscans in Tetouan as a young girl, joined the Municipal Pharmaceutical Inspector Corps in 1954, shortly before taking the position of assistant professor at the Universidad de Madrid, working under Professor Fernando Burriel (Arribas Jimeno, 1985: 39–41).

This trend continued. In 1962, the name of Concepción Ramos, a graduate in Physical sciences, appeared after her co-author of the article on descreening ('Destramado', 1962), José Pastor.[26] For her part, A. Carbonell, also a graduate in physical sciences and fellow in the Crystalline Thermodynamics Section, saw her name between those of José Luis Amorós, the head of the Department of Physical Crystallography, and his wife, the scientific collaborator María Luisa Canut in their work on an X-ray diffractometer ('Un diafractómetro de rayos X de dos limbos para altas y bajas temperaturas y su utilización en medidas absolutas de difracción difusa, 1962), a study funded in part by the 'US Department of Army' through its European Research Office, contract number DA, 91-591-EUC-1717.[27] Interestingly, María Luisa Canut also used the generic masculine gender to report her rank and institutional affiliation, which became '*Jefe de Sección*' (male section head) just a year later.

Finally, in 1963, Luisa María Valdés, a fellow in the same section as Carbonell, appeared after María Luisa Canut, but before department head José Luis Amorós, in the byline of an article on diffuse X-ray diffraction ('Empleo de análogos ópticos en el estudio de la difracción difusa de rayos X por los cristales',

1963), which was partially financed by the 'Air Force Office of Scientific Research OAR, Grant N.º AF EOAR 62-92' through its 'European Office, Aerospace Research, United States Air Force'. The article, however, was in fact 'an advance look at the doctoral thesis' by researcher Luisa María Valdés,[28] who read her work, directed by José Luis Amorós, at the Universidad de Madrid that year.

Leadership Positions: The Importance of Names

Although the *Revista de Ciencia Aplicada* was first published in the second half of the 1940s, it was not until 1955 that the name of any female author appeared in the journal associated with a high-ranking position. This was Piedad de la Cierva Viudes, a doctor in chemical sciences who was listed as head of the Chemistry Laboratory at the Laboratory and Research Workshop of the General Staff of the Navy (*Laboratorio y Taller de Investigación del Estado Mayor de la Armada*, LTIEMA). On that occasion, she was publishing her study of optical glass ('Ensayos de fabricación de vidrio óptico', 1955), which included teamwork from a 1954 project that had received the Juan de la Cierva Board for Technical Research Prize that year.[29] This pioneering woman – so often cited throughout the pages of this book – was also one of the first women to publish an article in the journal with only another female author, who had also been a researcher on her team and her doctoral student,[30] Francisca de Andrés, a doctor in chemical sciences and researcher at LTIEMA. Their article on Spanish clay ('Propiedades físico-químicas de arcillas españolas', 1958) opened issue 63 of the *Revista de Ciencia Aplicada* and was the third of a series of works they presented at international meetings and conferences.[31] The following year, Francisca de Andrés Contreras also appeared in the journal as a sole author of an article on optical glass ('Fabricación de crisoles para vidrio óptico', 1959),[32] again the third of a series of studies (the first two having been published in *Tercera Reunión Internacional sobre Radioactividad de los Sólidos*).

In the 1960s, female authors continued to appear in leadership positions throughout the pages of the journal *Revista de Ciencia Aplicada*. The November–December 1963 issue contains a work written by the head of the spectral analysis section, Carmen Álvarez Herrero, along with Amalia Cabrera, a doctor in chemical sciences, and Rafael Soriano Aroca, head of the Raw Materials Laboratory in the Valve Department of the company Marconi Española S.A. In 1964, seven years after her first piece, the doctor in Chemical Sciences Aurora Sampedro Piñeiro (?–1998) published in the journal again, now identified as the laboratory head of the spectrochemical section of the Antonio de Gregorio Rocasolano Institute for Physical Chemistry and the Central Laboratory of the national Spanish railway network.[33] However, as director of the CSIC Department of Analytical Chemistry, Fernando Burriel Martí appeared as the main author of the articles he wrote in collaboration with Álvarez Herrero in 1965 and 1966.[34]

It would not be until 1966 that another woman, doctor in chemical sciences María Josefa Molera, would appear as the lead author of an article on gas chromatographs ('Cromatógrafo de gases', 1966)[35] that opened issue 112 of the journal as head researcher in the chemical kinetics section of the Rocasolano Institute for Physical Chemistry, along with her collaborator José Antonio García Dominguez and assistant Ángeles Couto. This pioneer in Spanish chromatography, collaborator and CSIC fellow – she would eventually rise to the position of research professor – also received funding in 1959 to study in the Physical Chemistry Department at the University of Sheffield under Professor G. B. Porter, the future 1968 Nobel Prize winner for chemistry.[36]

Decades after Julio Casares Gil rejected her as his doctoral student because she was a woman, Molera was already one of the female veterans who had attained some degree of power and prestige in an institution, the CSIC, that divided these privileges extremely unequally between men and women. By 1970, half of the women who joined the CSIC staff in the 1940s had already reached the highest position in the council, that of research professor. Of the group that entered the council in the 1950s, however, only 28 per cent reached that level. This trend made it clear that, in equal conditions, seniority did not confer upon women either the prestige or the power that their male colleagues enjoyed (Alcalá Cortijo and Magallón Portolés, 2008: 161).

Something similar might be said regarding the female authors in the journal *Revista de Ciencia Aplicada*. In short, of the 277 authors who wrote – either as sole authors or co-authors – the articles published in the *Revista de Ciencia Aplicada* during the twenty-year period between 1947 and 1967, some thirty were women.[37] All of them had some type of connection to an official institution, usually as fellowship holders. This contrasts with most of the male authors, of whom only thirty-three were fellows (three former fellowship holders), while the rest were engineers either working in Spanish or foreign universities as professors or chair professors; at the Spanish National Research Council itself, primarily as directors, subdirectors, advisors, section heads, collaborators or researchers; at LTIEMA; or at companies and institutions such as Marconi Española S.A., the Spanish Electricity Industry Association (*Unidad Eléctrica S.A.*, UNESA) or the National Currency and Stamp Manufacturer (*Fábrica Nacional de Moneda y Timbre*, FNMT), all recognised positions that provided the authors with symbolic capital.

Part III

Printed Objects

Chapter 9

A Format for Science: The CSIC Technical and Scientific Journals

In 1952, the Publications Office of the Spanish National Research Council published a red and green brochure containing more than thirty pages of information – with illustrations, maps and photographs – about the founding and aims of the CSIC, its governing and research bodies, its libraries, publications, fellowship exchanges and research awards. This was one of the many propaganda brochures published by the CSIC in a number of languages for subsequent dissemination amongst official Spanish and foreign institutions. However, a large part of this information was also made available to the general public in the form of other, less serious brochures, like the one written by the CSIC secretary and poet Octavio Díaz-Pines y Fernández Pacheco – whose background made him the ideal person to direct the council's bibliographical newsletter, *Boletín Bibliográfico* in the early 1960s – part of the *Temas Españoles* collection on topics of interest to Spaniards, which could be found for sale in newsstands across Spain beginning in 1954.

However, as early as January 1954, the council's aforementioned institutional brochure was already almost sold out. In fact, Juan Roger, the head of the CSIC's Scientific Documentation Service at the time, wrote to Rafael de Balbín Lucas (1910–78), head of the CSIC Publications Office, to inform him that he had already spoken to José María Albareda about a new, updated print run of 5,000 copies. The new edition, he said, would be printed on 'much thinner paper to reduce mailing costs abroad', paper similar to that used for the most recent edition of the *Manual de Estudios* published by the Institute of Hispanic Culture,[1] and would also include some modifications, not only regarding the information on the CSIC boards and institutes contained in the previous brochure, but also international exchanges and the council publications, where 'it would be convenient . . . to add a reference to the titles of the books published in 1951, 1952 and 1953 to complete our list'.[2]

Table III.1: Books published by the Spanish National Research Council (1940–60)

1940–1	24
1942	69
1943	64
1944	90
1945	100
1946	168
1947	163
1948	179
1949	125
1950	192
1951	172
1952	138
1953	125
1954	96
1955	84
1956	89
1957	77
1958	104

1959	137
1960	118
Total	2,314

Source: Publications Distribution Section, Book Sales; AGUN/JIM/592/5

On 26 March 1954, the Sucesores de Rivadeneyra print shop sent these instructions to its workshop 'for prompt implementation'; however, seven months later, and after six galley proofs, the CSIC Publications Office had not yet received the copies, which posed a problem for the national and, particularly, international distribution of the brochure.[3] The ultimate aim of the brochure was to promote the work being done by the CSIC among official institutions around the world, informing them – according to this publication – that the council had a complex in Madrid made up of twenty-one buildings, including museums, institutes, archives and living quarters, with more than 1,600 published books 'on all the learned disciplines, from the physical to the biological sciences, from the biological to the spiritual sciences' and more than 150 published journals.[4] The numbers given for the books and journals need to be treated with some caution, as attested by the 1956–7 draft budgetary report, which, while asserting that 'the quality of life of the Council is reflected in its publications', revealed that during those years, the CSIC's printed materials had been affected by the council's budgetary deficit, with journals appearing on an irregular basis and the original publication and reprints of individual works deficient, all of which was unacceptable for a rigorous research organisation.

Table III.2: Council awards as of 31 May 1955

Year	Awarded	Published	Pending publication	On hold	Returned to author	No record
1940–1	7	7	-	-	-	-
1942	4	3	-	-	-	1

1943	10	9	-	-	-	1
1944	9	4	1	-	-	4
1945	17	13	1	1	-	2
1946	14	9	1	1	3	-
1947	14	10	2	1	1	-
1948	16	14	-	-	2	-
1949	15	11	3	-	1	-
1950	15	5	1	9	-	-
1951	16	4	3	4	5	-
1952	16	1	1	11	3	-
1953	15	-	1	13	1	-
1954	15	-	-	7	-	8
Total	183	90	14	47	16	16

Source: Publications Distribution Section, Book Sales; AGUN/JIM/592/5

In order to 'explicitly contribute to such scientific rigour', fourteen years after the creation of the CSIC, some of the senior members of the council and the editorial team also recommended as essential 'that all the editions of the Council's books and the corresponding originals be obliged to contain the favourable report that unambiguously certifies that the works to be published are fully guaranteed scientifically'.[5] However, this tentative indication of a formalised peer review process (Csiszar, 2016) did not leave any trace in the editorial book files at this time. Therefore, with no information about the reviewers, and until further research uncovers new data, there is still some question about whether this recommendation was ever implemented or if the monographs and journals of the CSIC were ever subject to a system of outside evaluation during this period.

All things considered, if there is one printed format that historically has helped to demarcate scientific authority while progressively gaining influence as the primary textual medium for certified scientific knowledge, it is without a doubt the scientific journal. In the words of Ibáñez Martín speaking in the late 1940s, 'none of the disciplines being developed in Spain today, ably and assiduously, are deprived of this stimulating, flexible and effective tool of daily research work that we call the journal'.[6] Indeed, the CSIC president regularly referred to the contents of the council's journals to demonstrate, he said, 'the smooth, sweeping *crescendo* of scientific production underway, that of a youth that works and believes'.[7] This entire endogamic system of journals published by the CSIC, all containing works largely written in Spanish, performed an active role in shaping and maintaining the Spanish scientific community during the dictatorship (Canales, 2019; García Naharro, 2020b; Malet, 2008).

It is for that reason that the CSIC journals dominate the upcoming pages. They will also serve as the basis for providing additional details about production and circulation processes, always interacting with the applicable legislation and the official image of science championed by the Franco regime, an engagement that influenced the journals in the making. One such example could have been *Síntesis*, a periodical project that never saw the light of day (Díaz Hernández, 2008: 33–6), but whose organisers, Rafael Calvo Serer, Raimon Panikkar and Ramón Roquer, would become key figures in the development of the influential CSIC journal, *Arbor. Revista General del Consejo Superior de Investigaciones Científicas*. The first bimonthly, 24.5 cm x 17.5 cm issue appeared in January 1944 and had an original print run no higher than 2,000 copies, although this later rose to 3,000. The research and culture journal's first promoter and editor was Fray José López Ortiz (1898–1992), a former intern at the Board for Advanced Studies and Scientific Research (JAE) in Germany (Díaz Hernández, 2015: 17) and friend of Josemaría Escrivá de Balaguer, whose thesis he directed in 1939 (Pallol, 2014: 605). After 1942, he worked as chair professor in Church history at the Universidad de Madrid and vice president of the CSIC, and was appointed bishop of Tuy in September 1944. After accepting the bishopric, López Ortiz left his position as chief editor of the journal and was replaced by José María Sánchez de Muniain (1909–81).

As noted in the early 2000s by chemist and CSIC Director of Publications (1991–5) Alberto Sánchez Álvarez-Insúa (1942–2011), 'Since its founding, *Arbor* was always controlled by the National Catholic postulates that guided the CSIC, as epitomised above all by José María Albareda and Rafael Calvo Serer' (Sánchez Álvarez-Insúa, 2007: 279–81). Rafael Calvo Serer (1916–88), who joined the journal as its secretary in October 1946, argued with Albareda about whether the format of the blue pages that contained cultural information from abroad should be published as an offprint or integrated into the journal (in the end, Albareda won out and it was published as part of the journal). After 1948, when Florentino Pérez Embid (1918–74) became the secretary, the journal began to be published on a monthly basis. It also took a more cultural turn – publishing an Arts and Literature supplement from February 1950 to December 1960 with 'poems, stories, reports on music, cinematography and theatre, along with reviews of books and the news – and a markedly political turn in line with the intellectual pirouette of the third force (*Tercera Fuerza*) – a reconceptualisation of Franco's dictatorship against the Falange and the old regime Catholics – by Calvo Serer, the editor of *Arbor* from 1951 to 1953, when he was dismissed from his posts at the CSIC as a result of his political positions (García Naharro, 2015b: 772).

With José María Otero de Navascués as chief editor (1953–6), *Arbor* continued to be a monthly publication with the same sections from before – 'Studies', 'Notes', 'Cultural Information from Abroad', 'Cultural Information from Spain' and 'Bibliography' – although the subjects covered tended to be more scientific and technological, an orientation that continued under the direction of José Ibáñez Martín (1956–69). However, although the journal pressed on with its campaign to disseminate the work being done by the various CSIC institutes, it took advantage of opportunities to pay homage to figures like Gregorio Marañón and Pope Pius XII and to provide laudatory reviews of events like the Second Vatican Council and the twenty-fifth anniversary of the founding of the CSIC itself. In turn, as an important official publication of the council, *Arbor* also attracted publicity in Spanish journals and newspapers, and was subscribed to by the country's ministerial bodies (Díaz Hernández, 2015: 24 and 42). In short, *Arbor* was always a well-connected journal that kept up with the times.

In early 1960, under the direction of Ibáñez Martín (now Spanish ambassador to Portugal), *Arbor* had an annual subscription price of 160 pesetas, which, according to Rafael de Balbín, needed to be raised in order to compensate for the difference between the journal's expenses and revenue.[8] One year later, the editorial board considered the possibility of making the July–August journal a special issue, devoted to 'economic and industrial development in Spain to celebrate two years of the stabilization [plan]'.[9] To that end, on 2 October 1961, José María Mohedano, secretary of *Arbor*, sent a letter to Ibáñez Martín, expressly asking him to contact a series of 'public figures' and request either a grant or a commitment to acquire a significant number of copies of the monographic issue being prepared as a way, Mohedano said, to help 'defray the higher costs associated with printing this special issue'. He added that 'given the considerable influence that you have with these public figures, it will not be difficult to secure these grants or commitments'; indeed, three years earlier, Ibáñez Martín had 'obtained subsidies using this same procedure' for the publication of a special issue dedicated to the fundamental principles of the Falangist Movement.[10]

In response, Ibáñez Martín wrote a series of letters dated 22 October 1961, requesting express support 'for the journal *Arbor*', saying that:

> This involves an extraordinary effort and it will be very difficult for us to achieve with our modest resources, and for that reason, I am contacting you to ask that you help us with some grant or, at the least, to commit to acquire a significant number of copies, a collaboration that will allow us to accomplish this very interesting enterprise elegantly. I appreciate anything that you can do in this respect. Kindest wishes from your friend, José Ibáñez-Martín.[11]

These letters were sent to the prominent figures proposed by José María Mohedano: Undersecretary of the Presidency Luis

Carrero Blanco, Minister of Agriculture Cirilo Cánovas García, Minister of Finance Mariano Navarro Rubio, Secretary General of the Movement and National Delegate of the Spanish Syndical Organisation José Solís Ruiz, President of the National Economic Council Pedro Gual Villalbí, Minister of Industry Joaquín Planell, Minister of Foreign Affairs Fernando María Castiella, President of the National Institute of Industry Juan Antonio Suanzes, Governor of the Bank of Spain Joaquín Benjumea, and President of the Governing Board of the Chambers of Commerce, Industry and Navigation Alfredo Mahou. Clearly, not just any journal was able to count on the social capital of José Ibáñez Martín. This, then, was one of the techniques used to promote and cover the costs of a journal that at no point abandoned its ideological hallmark in any way. In this regard, writing to Eduardo García Corredera (a member of the *Arbor* editorial board) after receiving the galley proofs of the articles that would comprise the special issue, Ibáñez Martín said:

> I think that more importance should be given to the years of the Franco Regime. I assume that you have read all the articles and are certain that their publication is not going to cause disturbances of any kind. I likewise believe that the most essential parts have been seen by the respective Ministers or their representatives, a guarantee that will help us to avoid all sorts of trouble. If that is the case, I expect that the issue will be a true success for *ARBOR*.[12]

To which Eduardo García Corredera responded:

> Although the issue says some things, its tenor is generally discreet (in my humble opinion), as the authors are attached to official organizations. I expressed this to Mr Mohedano, who apparently had specific orders from you and he soothed my conscience with respect to the censure of the contents. We expect the issue to be a success and that it will serve the current political moment in Spain well.[13]

Thus, in light of the politicisation of *Arbor*, scientific work had no problem finding its place in the many periodicals published by the various CSIC institutes. All these journals and books were regularly mentioned in the CSIC *Boletín Bibliográfico*, which was created in 1944 'to acquaint scholars with the authorised bibliographical material published by the Spanish National Research Council under the noble patronage of Generalissimo Franco, Caudillo of Spain'.[14] This eclectic catalogue listed a wide range of journals concerning 'spirit, life and matter' (García Naharro, 2017a: 245–62) including, since early on, numerous scientific and technical journals[15] with subjects that, in 1954, were divided between the biological, geological and exact, physical, chemical and applied sciences.[16] Examples include the weekly journal *Farmacognosia*, published by the José Celestino Mutis Institute, whose annual subscription price in 1944 was 50 pesetas (decreasing to 40 pesetas in 1946); the annual journal *Anales del Instituto de Edafología, Ecología y Fisiología*, containing monographic works, available in 1944 for an annual subscription cost of 25 pesetas or 15 pesetas for an individual issue; the monthly publication *Archivos de la Sociedad Oftalmológica Hispano-Americana*, priced at 60 pesetas and offering pieces written by doctors, naturalists, physicists and chemists; the quarterly *Trabajos del Instituto Cajal de Investigaciones Biológicas*, with an annual subscription fee of 50 pesetas in 1945; and *Anales de la Real Sociedad Española de Física y Química*, then the official, monthly publication of the Alonso Barba and Alonso de Santa Cruz Institutes, which cost 80 pesetas a year for a subscription. During the first third of the twentieth century, this outstanding journal was practically the only Spanish specialist publication on its subject.

Although other specialist publications could be found during the Franco period on, for example, optics, nuclear physics, metallurgy and electronics, *Anales de la Real Sociedad Española de Física y Química* continued to be the most important journal in the field (Valera Candel and López Fernández, 2001: 16–19), still providing the society with an outlet for its work. The Spanish Society of Physics and Chemistry (SEFQ), founded on 23 January 1903 after a meeting with the dean at the Universidad Central Faculty of Sciences, constituted its governing board two months later, led at the time by José Echegaray (1832–1916), who was also president of the Academy of Sciences in Madrid. That same year, the official SEFQ journal began to publish works that had been previously presented at regular meetings – thus providing the authors with public exposure – alongside news related to the society, like the proceedings of meetings, annual reports, income and expenses and an index of journals and books received by the organisation. In 1912, two major sections were introduced, one on original reports and notes and a second containing extracts and reviews, with summaries of articles published in international periodicals. Sixteen years later, in 1928, 'given the virtual absence of any relationship between the scientific and industrial worlds', the society proposed the creation of a technical section to rouse the interest of industrialists in scientific research and its technical application. However, this initiative only continued for two years, at which point these types of article were integrated into the Original Reports and Notes section (Valera Candel and López Fernández, 2001: 35–6, note 37).

Although the society was underwritten by the fees of its members, from 1910 to 1936 it also received financing from the state, which covered between 20 and 40 per cent of the society's total budget. Like the Spanish Society of Natural History and Spanish Mathematical Society, being a private enterprise made it easier for the SEFQ to continue to operate during the Franco regime after the uncertain period of the Spanish Civil War. In fact, on 9 June 1939, the journal began to operate again, once more with the designation 'Royal'.[17] However, the society – as was the case with other official and semi-official groups – was purged of any members who had been condemned for actions against the 'Glorious Movement', separated or ousted from any state body or official body or had simply 'collaborated in the scientific work of the Society between 18 July 1936 and 28 March of the Current Year of Victory'.[18]

In May 1940, a new issue of *Anales de la Real Sociedad Española de Física y Química* appeared, beginning a decade of marked government interference. For example, specific researchers were forbidden from publishing, due to their lack of 'ideological soundness' (Valera Candel and López Fernández, 2001: 208–12). In practice, this contradicted with the part of the official rhetoric of the regime which, since the second half of the 1940s, strove to confirm that 'no extant bud of collaboration was uprooted and no researcher denied the cordial opportunity to add their contribution to the collective enterprise'.[19] Once more, the tension between discourse and practice reared its head. Indeed, a comparison of the names of the most prolific pre-war and post-war authors reveals that of the forty-nine authors with more than two articles published in *Anales de la Sociedad Española de Física y Química*, only five published both before (1931–7) and after (1940–9) the Spanish Civil War. Moreover, although the number of articles did not decline, between 1911 and 1937, almost 75 per cent – in this case, on the topic of physics – came from authors with ties to the Board for Advanced Studies and Scientific Research (JAE), while after the Civil War, most of the writers were working for the CSIC or the university.

The royal society (RSEFQ), then, was now institutionally linked to the CSIC, becoming the 'communication channel for up to 5 Council Institutes: Alonso Barba (Chemistry) and Alonso de Santa Cruz (Physics) and Daza Valdés (Optics), Gregorio Rocasolano (Chemistry-Physics) and Electricity and Automation', providing a fixed financial compensation and variable publication quota according to the number of articles appearing on its pages from each institute.[20] Additionally, during times of scarcity, the RSEFQ supplied paper for CSIC institute publications, as in the case of the National Geophysics Institute, which requested paper to publish its offprints in 1941 and received it from the reserve fund set aside by the society for these institutes, 'under the condition that they return what is needed for the works that this society has pending publication'.[21]

All the CSIC institutes enjoyed the privilege of never having to cease publication. The National Institute of Medical Sciences produced monographs containing more than 400 17 cm x 24 cm pages for 40 pesetas in 1945; the Royal Botanical Garden had its *Anales del Jardín Botánico*; the Spanish Institute for Entomology published an eponymous quarterly journal for 9 pesetas per single issue and 32 pesetas for an annual subscription in 1948 (which rose to 10 and 38 pesetas one year later); and the Lucas Mallada

Institute issued *Estudios Geológicos*, which reflected not only the institute's endeavours, 'but also the many individual specialists who deserve to be disseminated, as well as the scientific movement in general that influences Geology, with an important bibliographical selection of reviews that increase the usefulness of this publication', and was available in 1950 for 25 pesetas for a single issue or 40 pesetas for an annual two-issue subscription, upon sending a request to the CSIC Publications Office, now located on Vitruvio Street in Madrid (López Sánchez and Fernández Gallego, 2021).

Around this same time, the Institute for Statistical Research also published its journal, *Trabajos de estadística y de investigación operativa*, a 24 x 17 cm publication with a print run of 600 copies that included research work, notes and a bibliography of interest. This format was also used for the quarterly *Revista de Geofísica*, with a print run of 535 issues, published and supplied by the National Institute of Geophysics at a cost of 12 pesetas for a single issue and 35 pesetas for an annual subscription.[22] Beginning in 1947, the Jaime Ferrán Institute for Microbiology published its quarterly journal *Microbiología española* for 22 pesetas per issue or 80 pesetas for a subscription. The quarterly *Genética Ibérica* was available for the same price and, like *Farmacognosia*, was published by the José Celestino Mutis Institute for Pharmacognosy.

Between 1960 and 1978, the journal *Luz*, published by the Daza de Valdés Optics Institute, had a significant print run of 1,500 copies, with a 22 cm x 28 cm format. However, one of the centres with the greatest editorial production was the Jorge Juan Mathematics Institute, which in addition to the construction journals *Estructuras*, *Gaceta Matemática* and *Revista Matemática Hispanoamericana*, had a variety of book collections: Mathematics Conferences, History, Philosophy and Mathematical Methodology, Mathematics Reports of the Jorge Juan Institute, Mathematics Monographs, Classic Mathematics Texts and Modern Mathematics Texts (García Naharro, 2015b: 772–4). However, for the purposes of this book, the most active centre by far, both in terms of its activities associated with construction and scientific and technical research (Garma and Sánchez Ron, 1989: 60) and with its editorial production, was the CSIC Juan de la Cierva Board, which published a vast number of influential journals on subjects ranging from general construction questions to specialist publications on pre-stressed concrete, grease and oil, plastics, welding and more. What follows is, precisely, a detailed study of many of these journals, conducted in an attempt to understand how they took shape and what role they played, questioning them not only as texts *per se*, but as products in process and in transit which, in turn, convey information of interest about their status and how they were used. Particular focus is paid not only to their materiality – their forms, contents, sections and illustrations – but also to their value as artefacts in circulation, looking at both the places and points of sale and the frequency of their publication. However, the real objective here is to complete this analysis of frequencies and social ties with a consideration of 'competences and uses' (Bourdieu and Chartier, 2010: 259). This alone is the ultimate motivation for comprehending the relationships established between the journals, understood as 'printed objects', and their associated value, as well as examining the characteristics and profiles of their readers and / or consumers.

Chapter 10

The Journal as Object of Study: Epistemological Concerns

Scientific journals are socially identified as the place into which scientific communities pour their knowledge, a public – and published – space where the interested reader can expect to find in-depth articles by writers considered qualified experts in various areas of thought. These modern periodicals really began to appear in the nineteenth century as a new platform for scientific authority, whence they helped to delineate the space of worthwhile scientific knowledge. However, this space had not always been so well defined; the attributes of the old, primitive scientific journals of the seventeenth century were not so marked and they did not enjoy the social consideration that the format began receive in the mid to late nineteenth century, when it gradually emerged as the legitimate, institutional locus of scientific authority (Csiszar, 2018; Houghton, 1975: 11–31).

'Learned journals contributed significantly to a sharper definition of the professional scientist', wrote Elizabeth L. Eisenstein, 'to new divisions of intellectual labour and to the creation of a "referee system"' (Eisenstein, 1990: 16). As an editorial product, the pages of the scientific journal united an author with a discourse and a collection of earlier references that validated a discovery or a postulate that, once enunciated, became fixed, inscribed in time and in a textual space that certified the material device of the publication itself. To be understood as such, scientific journals must meet a series of formal requirements, including primarily consisting of scientific articles signed by their authors, who take the credit and the responsibility for the assertions in the text. Moreover, their assertions are built on prior research to which the text refers in bibliographical references and citations (Csiszar, 2018: 1–21; Delgado López-Cozar and Ruiz Pérez, 2009).

Using a critical apparatus comprising authoritative references, the scholar interested in the subject can analyse how scientists accommodate the enunciations of others to construct the arguments that shape their own scientific articles, studying, for example, how the enunciator critiques, challenges or adheres to the bibliographical or formal sources: direct citation, indirect citation, paraphrase, repudiation (Murga, 2005: 11–18 and 57–78). Other supporting elements like charts and figures may complement and reinforce the text: tables, equations and captions that, when, added to the rhetoric and external references, stratify the texts, endowing them with a density of highly interesting information (Halliday and Martin, 1993; Latour, 1992: 21–62).

For this reason, when working with the scientific text, one must be aware of what the simple formal and discursive structure implies. Textually speaking, the literary techniques characteristic of specialist articles point towards an ideal type of audience primarily situated in scientific and academic circles; due to their subject and the technical nature of their language, these articles are not easily accessible to a lay public (Latour and Woolgar, 1986: 43–104). However, while based on these premises, the following pages focus on analysing the social status of the textual device – in other words, the journals themselves – paying attention as well to the places where they could be acquired and the institutions that gave them their backing. In this case, these elements were determining factors when selecting the readers – and the ways in which these journals were, or could be, read – under a dictatorship.

To that end, these chapters analyse the formal appearance of a variety of different journals, all official periodicals of the Spanish National Research Council, printed products articulated in a context of socially shared expectations and knowledge that, in turn, prescribed the specific ways they could be employed by their users. In their construction, then, the journals are measured by socially established conceptions of what scientific journals are, what they should be and to whom they should be directed (Woolgar, 1991: 57–102). In this case, the scientific and technical journals were produced by the Juan de la Cierva Board for Technical Research, the primary technical and industrial research centre during the Franco period.

With respect to the CSIC, the Juan de la Cierva Board had its own legal personality and relative financial and administrative autonomy, and the body was tasked with 'research missions of a technical and industrial nature, subject to the fundamental guidelines of the unity of science in the service of the national interest'.[1] These official, regulatory assignments also applied to the board's technological studies, bibliographical works, research areas and patent exploitation, as well as its general publications and journals. Indeed, like the very first publications that emerged from the various CSIC institutes beginning in 1940, the scientific journals of the Juan de la Cierva Board had to adhere to strict formal protocols. Writing from the office of the secretariat-general in 1940, Alfred Sánchez Bella spelt this out very clearly:

> On the top part of the cover, the title 'Spanish National Research Council' should appear and below that, the name of the Board to which the Institute corresponds; then the title of the journal or publication, followed by 'edited by' or 'the official publication of' the name of the Institute . . . On the inside page, there will be a portrait of Franco at the top with the following caption: 'To the Caudillo of Spain, under whose high Patronage the Spanish National Research Council was created'; on the following page you may add a photograph of the Minister with the inscription: 'Hon. José Ibáñez Martín, Minister of National Education, President of the Spanish National Research Council'. On the back cover and a bookplate, the 'Arbor Scientiae' . . . and this Secretariat must be sent 50 copies of the first publication, printed, if possible, on premium class paper.[2]

These publications, then, were subject to rules, but also enjoyed the resources and means at the disposal of a regime that promoted, distributed and disseminated its editions through official channels, while prioritising the supply of paper to the needs of the council. In fact, although paper was scarce for a number of years, it was never lacking as far as the CSIC publications were concerned. However, beginning in 1953, and based on 'the lessons learned in the CSIC Publications Office', the council recommended that the journal print runs be standardised to 'lower editing costs and economize on storage', fixing the print run as the sum of a) current subscriptions; b) exchanges; c) sales of single issues; d) 10 per cent of the sum of a, b and c; and e) 150 permanent collections.[3]

The high production costs of the journals, however, were at times simply due to the conditions surrounding their production; the excessively wide margins, inordinate line spacing and large type fonts used increased costs while giving 'the disturbing impression of a lack of editorial moderation, which is anomalous to the typographical customs of scientific publications'. Accordingly, a recommendation was made to establish the font size, the measurements for the typographical box on the printed page, as well as the number of lines per page, the line spacing and the number of matrices per line. In other cases, the high cost resulted from a disparity between the print run and the number of subscribers, a more pressing situation with the journals from the scientific branch of the CSIC, as seen in this letter to the science commission:

> In the best-case scenario, the print run is 500 copies and the number of subscribers 137. There are two other, similar cases: one with 800 printed copies and 137 subscribers, and another with 700 copies and 137 subscribers. These are followed by five journals with print runs of 500 copies and fewer than 100 subscribers. In the worst case, a new journal, a 1,000 copy print run that cost 44,000 pesetas sold 10 copies.[4]

While the conspicuous disproportion between the print run and the actual market needs was discussed behind closed doors as creating space problems for the Zabaleta warehouse where the journals were held, for outward appearances, the number of printed pages and print runs per year were used by the dictatorship as an unequivocal sign of the strength and apparently good health enjoyed by its research centres. This, however, did not reflect the reality of the situation, as recognised by the director of the Rocasolano Institute, Antonio Rius Miró, in November 1953. He shared his concerns, that all his work plans were in 'imminent danger of collapsing if the resources of the Institute [for physical chemistry, during 1954–55] did not substantially increase',[5] with the director of the Jorge Juan Institute for Mathematics, who then wrote a letter to President José Ibáñez Martín in which he brought up the council's alarming economic situation, saying:

> I take the liberty to disturb you because of a rumour that has reached me, that it is not going to be possible to increase the budget for this Institute for the coming year, catastrophic news for us. It was already impossible for us not to exceed the Institute's working budget in recent years, something reflected in its publications, which are in no way cheap and increasing quickly year by year. The unavoidable result is going to have to be stopping the presses, like was already done to us twice this year with the oldest one. And suspending the publications, especially the journals, represents a public confession of impotence both abroad and at home, beginning with turning our backs on our exchange commitments and our obligations to our subscribers. I do not need to stress to you the importance of the exchanges, which is almost the only possible way to obtain the extremely expensive foreign journals whose articles, as you well know, are the scientific researcher's primary tool, particularly in Mathematics. I earnestly beseech you to do whatever is within your power to avoid this disgrace.[6]

Although these fears reflected the changes that were taking place in the publication services of scientific societies around the world at the time (Fyfe, 2021), they were also firmly cemented in the specific conditions affecting some of the council's publications. For example, the biology journal of the Cajal Institute was suspended for months in 1956, and although the medical journal *Anales de Hidrología Médica* published 1,000 copies, only three of these were distributed, while, at the same time, the steering committees of other journals begged for 'the original of each issue'. In this situation, with 'journals reaching minimum subscription and newsstand numbers', he suggested that the large number of journals be decreased, as expressed by the president of the CSIC Scientific Publications Commission, Dr Gregorio Marañón. Reducing the excessive number of scientific periodicals would cut expenses, relieve storage problems and 'bring together journal readers with a similar scientific spirit,' in the words of Marañón, 'thus avoiding the asphyxiation of scientific thought produced by its own poorly controlled fecundity'.[7]

As Marañón's letter reveals, the vast number of journals published by the council institutes was only seen as a way to 'facilitate publication' for council researchers and exchange the journals for other, always expensive foreign periodicals. In fact, the CSIC dedicated a good portion of its editorial production to exchanges, donations or complimentary shipping – which were not even always well received – reserving, moreover, 'up to 10 per cent, at the most, of the book print runs and 10 copies of each journal issue' to be reviewed in specialist journals and the press.[8] As observed by National Institute of Industry president, Juan Antonio Suanzes, another fear – and a well-founded one – was that 'the findings would remain "frozen", only known to specialists or scholars who regularly read our journals and publications, without them coming to the notice of the companies or individuals that could better and more quickly use them in their applications':

> Please understand that this approach to the problem is not an attempt to deviate from the normal channels of publication through which the results of scientific research must always flow. On the contrary, this is why I must draw the attention of the Institute Directors and researchers in general to the greater international dissemination achieved by publishing the results of their work in journals, and not in monographs, which largely remain in our inner circle. By exchanging these journals, it becomes easy to disseminate extracts and, thus, Spain's contribution to scientific and technical production is held in higher esteem in every instance. The aim is not to view publications in themselves as the finishing touch of the

> work, but to meditate about what can be done so that interested parties and industries receive the results more quickly and easily.[9]

Supporting the journal – as the preferred format and official vehicle of the scientific publication system – to the detriment of monographs, and arguing on behalf of publishing these results in specialist journals – with schematic sketches in non-specialist publications – would, according to Suanzes, 'bring the results of the work of a few (in the world of research) to the initiation of applications of many (in the world of industry)'.[10] This supply of scientific and technical information to Spanish industry was one of the roles developed by the Juan de la Cierva Board, not only through the editorial products of its Information and Documentation Centre (*Centro de Información y Documentación*, CID) (Juretschke, 2001: 989–94), but also through the periodical publications themselves, which fulfilled this and other functions. Against this background, a selection of the journals and bulletins published by the various institutes connected to the Juan de la Cierva Board are discussed in greater detail in the next chapter. These, however, are all studied through the sections and representations within these printed artefacts; under Franco, of course, the very format and materiality of the official scientific journal was political.

Chapter 11

The Scientific and Technical Journals of the Juan de la Cierva Board (CSIC): A Whole Human Comedy behind the Periodicals[1]

Revista de Ciencia Aplicada (Journal of Applied Science): *Researchers and Industrialists as the Target Readership*

This dark red 21.5 cm x 29.5 cm paperback edition is dominated by its title, *Revista de Ciencia Aplicada,* written in large jet-black Roman capital letters. The journal's clear affiliation with the Spanish National Research Council appears directly below the name in smaller black capital letters: 'Published by the Juan de la Cierva Board for Technical and Scientific Research – (C.S. de I.C.)', below which, stamped in white – the same colour as the border framing the information on the cover – is the council's bookplate image: the *Arbor Scientiae.* The same emblem also appears stamped in black in the middle of the white back cover, accompanied only by its affiliations: 'Spanish National Research Council' and 'Juan de la Cierva Board for Technical and Scientific Research', respectively, above and below the CSIC stamp.

The *Revista de Ciencia Aplicada* maintained this sober and austere presentation even in the late 1960s, twenty years after it was founded. The bi-monthly publication, overseen by José Castañeda, was managed and edited in its permanent headquarters at 150 Serrano Street in Madrid, which also served as the outlet to buy a copy or request an annual subscription. Castañeda, a professor of economic theory in the Universidad de Madrid's Faculty of Commercial, Economic and Political Sciences, as well as head of education for the Legislation, Organisation and Economics

Group at the School of Industrial Engineers in Madrid,[2] served as editor-in-chief of the journal beginning in 1948.

Since 1946, the Juan de la Cierva Board had been looking into the possibility of publishing a 'Central Board Journal' that would contain 'articles on economic-technical problems, updates on technical-scientific topics, a patent review, refereed sections, general news about the board's activities, etc'.[3] However, the first, initially quarterly, issue was not published until late 1947, corresponding to October–December and costing 20 pesetas (25 pesetas for a back issue) or 75 pesetas for an annual subscription in Spain, 'the Moroccan Protectorate, the Colonies, Portugal, the Americas and the Philippines' (100 pesetas for the rest of the world). Initially, a subscription had to be requested at the former Juan de la Cierva Board headquarters at 95 Alcalá Street in Madrid, where the journal was also being edited and managed at the time.

The first sixty-four-page issue was printed – like all the issues of the journal over the years – by the Madrid printing house Blass, S. A. Tipográfica, located at 27 Núñez de Balboa Street. The front page was dedicated to a reproduction of an image of the head of state, Francisco Franco Bahamonde, above a caption that read, 'Under whose eminent patronage the Spanish National Research Council labours'; the board's new journal could not begin without paying homage to the sponsor of science in Spain. After this image, the journal was divided into sections that, with some slight alterations, would continue throughout its existence: first a series of five scientific articles, written by researchers whose institutional affiliations were included in a footnote, followed by 'Information from Abroad', 'Information from Spain' and 'Bibliography' – which, as the journal itself noted, contained 'critical references from works whose publishers sent two copies to the Editor-in-Chief of the *Revista de Ciencia Aplicada*'[4] – and ended with a bibliographic index of the books and leaflets, journals and 'technical journals received by the "Juan de la Cierva" Board'. The publication also offered a photocopy service, advertised on its pages, 'to be able to supply them to interested persons or companies'.[5]

The purpose of the journal, as described in the editorial titled 'Greetings', was:

> A three-pronged mission, to introduce the Board's activities in the broad field of applied science, to gather the latest news from the world of technology in its diverse branches and to provide researchers and industrialists with the necessary documents for their work, informing them through notes and bibliographic records about the principal books that have appeared and the contents and general ideas in the most notable articles, found in a vast array of specialist journals.

Individuals and enterprises – or more specifically, 'researchers' and 'industrialists' – were therefore, the prototype of the ideal target reader for the contents of this journal, although its first issue also contained a greeting 'in the best spirit to the general public' alongside its aim to improve, 'reduce weaknesses and iron out imperfections' to become, in its words, a journal 'worthy of the recognition of our readers and to appear both amongst the activities of the "Juan de la Cierva" Board and the ranks of the scientific press, both national and around the world'.[6]

To that end, the texts met the formal requirements preestablished for every self-respecting scientific journal: articles by authors connected to research centres, with abstracts introducing articles written using a technical language, always accompanied by footnotes, tables, charts, graphs, figures and diagrams, formulas, equations and numerical results. These elements complemented and reinforced the arguments made by the authors, assembled and interspersed through the body of texts that in many cases came with a specialised bibliography. As Juan Antonio Suanzes observed in 1949, 'The value of the authors, the structure and the careful composition justify the success of the endeavour'. That same year, the journal began to be published on a bimonthly basis, a change 'made possible and advisable because of the quantity and quality of the originals that it is receiving':[7]

> With the growing frequency of communication with the public, we intend to increase interest in our more up-to-date informative Sections, which have been so magnificently received by our readers, while maintaining the length of each of our issues, with the volume increasing by 50 per cent, allowing us to provide space for the expanding number of works from our authorized collaborators.[8]

The increase in the expected number was also accompanied by 'successive improvements in the quality of our pages', something that became quite clear after issue 13 (March–April 1950). As a bi-monthly publication, by 1952 the *Revista de Ciencia Aplicada* had 581 subscribers, was exchanging publications with sixty-three Spanish and 294 foreign journals and giving away a significant number of its issues at no cost. The exchanges both helped to fill the collections of the Juan de la Cierva Board library and to 'alleviate' – as it was said in 1960 – 'subscription costs',[9] which comprised its greatest expenditure by far.

Table III.3. Number of Subscribers

Year	**Number of Subscribers**
1950	256
1951	557
1952	581
1953	260
1954	No data
1955	328
1956	419
1957	427
1958	396
1959	389

1960	No data
1961	372
1962	No data
1963	312
1964	318
1965	326

Source: Juan de la Cierva Board for Technical Research, annual activity reports

Table III.4. Number of exchanges

Year	**Number of exchanges** Spanish journals/international journals
1949	39/46
1950	49/136
1951	51/265
1952	63/ 94
1953	65/295
1954	69/299
1955	73/307
1956	70/290
1957	70/353
1958	75/382
1959	82/406
1960	No data
1961	95/471
1962	No data

1963	147/518
1964	131/516
1965	146/507

Source: Juan de la Cierva Board for Technical Research, annual activity reports

As seen in the first table, the number of subscribers decreased dramatically in 1958, something that the CSIC tried to ascribe in the secretary-general's records to 'declines in overall subscriptions from certain Institutions, a consequence of a general criterion applied to a number of different journals, which meant they reduced their subscriptions from 50 to 20'. However, to soften the blow, they added that the journal was continuing its '382 foreign exchanges and 75 national exchanges that feed the Board library, with a slight increase around 1957'.[10]

In any case, by 1954 the *Revista de Ciencia Aplicada* had a distribution of 719 issues,[11] and one year later it added a new section, 'Scientific Activity of the Board', which presented 'summaries of the works published by the scientific staff of the Institutes'[12] in addition to the bibliographic section, with its 'summaries of some 480 article indices per issue',[13] which were first published in 1950 and by 1964 contained more than 2,600 cards for articles from other publications.[14] As the journal noted in 1950:

> In order to expand the possibilities of using the Juan de la Cierva Board for Technical Research Periodical Archives, we have replaced the references to articles with an article index in a universally accepted form; this has notably increased the number of articles listed in the space that our Journal is able to dedicate to this section.[15]

This formal modification was primarily aimed at making it easier for users of the board's periodical archives – ideally the members of the staff of the Juan de la Cierva Board – to take advantage of the library's resources and to remind them that, in addition to having the opportunity to consult the journals that contained these articles, 'the Board's services can provide photocopies of the reviewed articles to whomever wants one'.[16]

Informes de la Construcción (Construction Reports): *A Show of Force Based on Printing Paper, Translations, Advertisements and Tear-Out Index Cards*

The history of the Technical Institute for Construction and Cement (*Instituto Técnico de la Construcción y del Cemento*, ITCC) is unequivocally connected to the figure of its founder and director, Eduardo Torroja Miret, who died in his office on 15 June 1961 and after whom the institute was named one month later as a posthumous honour.[17] However, the germ of this organisation dates back to 1934, to the Institute for Construction and Building, the private brainchild of a group of architects and engineers who were interested in studying all things related to construction and its materials. Years later, in 1946, it came to form part of the Juan de la Cierva Board, and in 1949 became the ITCC, the result of the merger between the Institute for Construction and the Institute for Cement.[18]

The Institute for Cement had been created in March 1947, although it did not officially begin to operate until October of that year,[19] with the engineer Félix González serving as president of its administrative board. This council was made up of the directors of the Central Laboratory for Testing Construction Materials (*Laboratorio Central de Ensayos de Materiales de Construcción*, LCEMC) at the Barcelona Technical Research Special School of Engineers of Roads, Canals and Ports and the army engineers laboratory. It had no building of its own, but maintained provisional offices located on Gaztambide Street in Madrid and, with more projects than accomplishments, the institute operated in collaboration with the above laboratories before the merger, working almost exclusively on compiling bibliographic information on cement.[20] Finally, after the creation of the ITCC, the organisation left both those offices and the spaces located on the lower floors of 25 Ruiz de Alarcón Street in Madrid – the headquarters of the Institute for Construction since 1948 – and those on 58 Lista

Street, to take up residence at 47 Velázquez Street in July 1949.[21] However, efforts continued to find a more appropriate location for the new institute, which began to operate in the Costillares complex in 1953, with the new premises and facilities being officially inaugurated by Generalissimo Franco on 8 February 1958. According to the annual report:

> The building, which is property of the Institute, is constructed on a lot some 55,000 m^2, located in the pine groves of Nuestra Señora del Recuerdo, in Chamartín. It occupies a covered surface area of some 7,000 m^2, with an average of two storeys and basements where the Clerical Offices, Laboratories, Pilot Institutions and so forth are distributed.[22]

This building also held the Publications Office, which was responsible for publishing not only the journals, leaflets and books produced by the institute, but also products from its partners, and assembled 'a card index of the national firms that produced construction machinery and material' that contained 5,388 index cards in 1953.[23] The Publications Office was also responsible for *Informes de la Construcción*, the institute's monthly journal that, during the first year of its existence, 1948, published six issues with a total of 494 printed pages. A mere year later, it published ten issues, with a total of 18,500 copies and 2,775,500 printed pages.[24] It was also at this time, in 1949, that two bulletins first appeared: *Últimos avances en materiales de construcción*, 'dedicated entirely to informing cement manufacturers about the most important foreign innovations', and its counterpart on advances in building, *Últimos avances técnicos en edificación*, which published 230,000 pages in 1954.[25]

Clearly, during the early years, there was a keen interest in entering the number of printed pages into the accounts: 'This Institute has therefore published 4,004,000 pages this year,' noted the 1949 annual report, 'in other words, a four-fold increase with respect to the previous year'.[26] This show of force by the institute was based on printing paper, a tradition that manifested itself in the following annual report as well, which provided information about the quantitative advances in the final total number of pages in its publications:

- *Informes de la Construcción*: in 1950 another ten issues appeared, with a total of 29,500 copies and 3,736,100 printed pages.
- *Últimos avances en materiales de construcción*: nine issues in 1950, with a total of 316,000 printed pages.
- *Últimos avances técnicos en edificación*: four issues in 1950, with 172,500 published pages.
- *Hormigón pretensado* five issues in 1950, 250,000 printed pages.

What established *Informes de la Construcción* as a consequential publication was not only the spectacular increase in its print run, but also the fact that it was the 'only'[27] scientific and technical journal whose articles were required to be published in three languages: French, English and German (Cassinello Plaza, 2008: 21). Boasting a colour cover after 1949 (the first issue was illustrated with the council's *Arbor Scientiae* logo), its careful design and inside pages, which were adorned with drawings and patches of colour, was one of the hallmarks. Against *Informes de la Construcción,* the earlier publications of the Technical Institute for Construction and Cement (twenty-six monthly issues of *Hormigón y Acero* and the annual ITCC reports published between 1941 and 1946, *Anales del Instituto Técnico de la Construcción y la Edificación*) paled in comparison, both quantitatively and qualitatively. Indeed, this was probably the most important journal in the field at the time in Spain, only sharing the spotlight with *Cemento Hormigón* (1929) and the Agromán construction company's publication, *Obras* (1931–86) (Cassinello Plaza, 2008: 5–6).

Founded in 1947, the first issue of *Informes de la Construcción* was published in May 1948 with a clear purpose: to inform technicians about the technical and scientific projects and advances of interest in construction coming from other countries. To that end, the journal studied foreign publications, summarising and adapting them for its Spanish readership, 'redoing figures and application examples when necessary, translating the figures that appear in the original article to the metric system and sometimes commenting on the work or adding explanatory notes'.[28] However, this was done without disparaging Spanish products. *Informes de la Construcción* also contained commercial information that, according to the journal, 'is published at no cost, for the sole purpose of supplying technicians with data about the materials currently available on the national market'.[29] In 1948, in fact, the journal agreed to publish 'articles containing technical-commercial information provided by companies, as long as they offer some significant innovation in the field of technology and are justified by their technical and scientific interest'. However, such articles were only published once, so as to prevent the journal from being used 'for propaganda purposes'.[30]

It was not until issue 83 (August–September 1956) that, under editor Fernando Cassinello, the journal began to include publicity. Guided by the Publications Office, *Informes de la Construcción* launched 'a campaign to disseminate all elements related to construction, whether Construction Companies, Machinery, Materials and Construction Materials, etc.' in the form of full-page advertisements. The institute reserved the right to edit and select them, and they were appropriately interspersed among the articles (Cassinello Plaza, 2008: 85–6). As a result, the number of advertisements in each journal grew exponentially from the original five in the 1956 issue, accentuated by occasional notices in different languages and colours that appeared under the caption 'Report on Reports'.

The advertisements were integrated intentionally, either immediately before or after a specific work that had used the material, technique or patent being advertised (Cassinello Plaza, 2008: 87–8). The aim was to provide information about innovations being developed in the field, playing with the space occupied by the advertisement and the design and layout, which was always supervised by the journal's editorial team:

> The journal designed the advertisements for its 'customers' at no cost, following up with each company and product, requesting information on a regular basis to keep the project images up to date,

> and connecting them to awards, competitions or any other type of activity in which they were involved at all times. (Cassinello Plaza, 2008: 90)

Like any self-respecting scientific journal, *Informes de la Construcción* was also structured around research articles that were divided into ten thematic sections: general topics; architecture and constructions; basic sciences and secondary techniques; auxiliary facilities; material resistance and structural calculation; engineering and constructions; material updates; price, regulation and test organisation; construction elements; and construction processes. These were usually written by ITCC staff or provided by foreign journals and translated into Spanish by council staff (the names of these 'other authors' – the translators – were not included as writers; at the most, their initials appeared alongside the article).

These articles were accompanied by brief notes, summaries of institute publications and activities, calls for competitions and bibliographic references, both national and international; this last section ended up taking on a life of its own, being published separately from the journal. These features were supplemented by removable 'question cards' with the address of the institute pre-printed for easy mailing, and 'special pullouts' with titles like 'Did you know that . . .' and or 'Have you considered that?', which gave advice or information about specific matters, usually practical in nature, as they were also the offprints of technical suitability documents (Cassinello Plaza, 2008: 128–30). What was most remarkable about this publication, however, was the vast quantity of collectable, tear-out index cards, arranged by subject, colour and code numbers, that came with the issues from May 1949 (No. 11) to September 1960 (No. 133):

> There were four types of cards: industrial, bibliographic, machinery and Institute calculation service. All these cards were inserted amongst the last pages of the journal, and were easy to tear out and catalogue . . . like the articles, these cards were coloured to make them easier to organize and recognize 'at a glance'. Issue 11 of the journal

> (May 1949) published the colour code under the title 'Classification Guide': White, code zero, General; Sea Green, code 1, Architecture and its Constructions; Red, code 2, Basic Sciences and Supplementary Techniques; Grey, code 3, Auxiliary Facilities in the Building and City; Sienna, code 4, Material Resistance and Structure Calculation; Blue, code 5, Engineering and its Constructions; Ochre, code 6, Knowledge of Materials; Orange, code 7, Regulations and Price Organization; Yellow, code 8, Construction Elements and Prices; and Greenish-Brown, code nine, Miscellaneous. (Cassinello Plaza, 2008: 126)

From the beginning, the journal could be found in the institute library, which had a collection of 643 volumes and received ninety journals on a regular basis in 1948, both by direct subscription and through an exchange system.[31] The exchange of publications, again, was a vital practice for the institute, as represented by the corresponding data from the 1950s:

- 1952: *Informes de la Construcción*, 1,524 copies; *Monografías*, 1,046 copies. In exchange for these shipments, they received nearly 2,500 various publications, including books, journals, offprints, etc., from a variety of sources.[32]
- 1953: *Informes de la Construcción*, 1,118 copies; *Monografías*, 525 copies; receiving in exchange more than 2,000 publications from 'a variety of sources'.[33]
- 1954: *Informes de la Construcción*, 1,939 copies; *Monografías*, 950 copies; receiving in exchange 2,200 publications.[34]
- 1955: *Informes de la Construcción*, 1,920 copies; *Monografías*, 1,615 copies; receiving in exchange 2,500 publications 'from a wide variety of sources'.[35]

In addition to publishing its institutional journal, the ITCC, like the Institute for Cement before it, was also active in publishing books that shared the spotlight with its *Monografías*, a set of publications that received backing even before the institute was officially constituted and that published an average of slightly over ten books a year between 1949 and 1965, resulting in a total output of 297 monographs by 1971.[36]

Table III.5. Number of monographs published

Year	Number of monographs published
1948	10
1949	17
1950	13
1951	9
1952	7
1953	13
1954	15
1955	19
1956	10
1957	10
1958	8
1959	9
1960	No data
1961	9
1962	9
1963	8
1964	11
1965	12

Source: Juan de la Cierva Board for Technical Research, annual activity reports

Monographs were also published by the Spanish Society for Soil Mechanics and Foundations, a branch of the institute, as was the Spanish Association for Prestressed Concrete. While the latter organisation once again began to publish its bulletin to disseminate prestressed concrete techniques in 1945, the Spanish Society for Soil Mechanics and Foundations put out both a foreign information and a bibliographic bulletin, as well as a report, along with dispatches from various soil mechanics societies from other countries in order, they said, 'to introduce them to all their members'.[37] The group also published *Cuaderno de Precios sobre coste de obras* and in 1955, on the occasion of the 'Basic Course for Project Assistants', held between 11 May and 27 July of that year under the direction of Francisco Arredondo Verdú, even published a wide range of related works.[38]

Returning to the ITCC itself, the institute published a series of official manuals on subjects ranging from heating techniques to working with reinforced concrete.[39] These publications came to comprise the series belonging to the standards and manuals section, an arm of the Research Department that published seven works in 1961, along with five other books that year associated with the construction materials research series.[40] Finally, these were accompanied by leaflets with 'primers, bulletins, indices and information',[41] some technical and informative booklets that shared the spotlight with issues of the activities newsletter, *Boletín de Actividades y Publicaciones del Instituto*, and the publication of offprints. By 1965, four issues of the *Bulletin* had appeared from what was now the Eduardo Torroja Institute for Construction and Cement, in addition to offprints 'of 100 articles that appeared in the journal *Informes de Construcción*' and, among others, six books from the Standards and Manuals collection, twelve original monographs and a similar number of journal issues, both from the ITCC and its affiliates.[42]

Boletín Informativo del Instituto Nacional del Carbón (National Coal Institute Newsletter): *Paratextual Combinations Portraying All Things Scientific as Self-Explanatory*

The *Boletín informativo* was initially a bi-monthly publication that became a quarterly beginning in May 1959, whose editor-in-chief was Francisco Pintado Fe. Its price per copy in 1955 was 20 pesetas (25 pesetas for back copies and 75 pesetas for an annual subscription) and the interested reader could acquire a copy by going to the centre itself, where the editorial staff and administration were also located: the National Coal Institute. Located in the town of La Corredoria in Oviedo in the Province of Asturias, in 1964 the institute was made up of a laboratory building and offices, 'with a surface area of some 2,200 m^2. It has three floors of laboratories (some 3,000 m^2 in total), three offices (1,320 m^2 in total), warehouses, a library and an assembly hall (under construction), etc. . . . There is also a shed for semi-industrial tests, some 700 m^2 in size'.[43] However, this had not always been the case; in 1947, the institute had neither staff nor working laboratories and it was through an agreement with the Universidad de Oviedo that the institute was able to begin its scientific endeavours in 1948.[44]

Shortly thereafter, in 1949, the institute published its first work, the first volume – the second was pending publication – on the origins, properties and use of coal: *El carbón: Algunos aspectos de los estudios sobre su origen, propiedades, preparación y utilización*. This book contained more than 350 17 cm x 24 cm pages, the same format as the work being prepared, which contained information about a series of conferences on soft coal held between August and September at the Universidad de Oviedo and was published in 1950. Similarly, in 1951, an illustrated paperback was published – this time by the National Fuel Institute – with the Spanish version of a conference arranged by Eugène Burstlein at the same university in December 1950.[45] Again, these editorial practices reveal how the CSIC institutes and the country's universities were, in practice, extremely close.

However, it was not until 1952 that the first copy of the *Boletín Informativo del Instituto Nacional del Carbón* came out, printed on 20 cm x 29.5 cm paper. In the middle of its sober black cover appeared the white letters INSTITUTO NACIONAL DEL CARBON, while the back cover was illustrated with the dark, circular emblem of the institute, which filled the middle of the cover, against a white background. Printed in Oviedo at the La Cruz printing offices (8 San Vicente Street), the *Boletín* contained articles, technical specifications and a list of the journals available at the institute library, as well as sections like 'News from our Institute' and 'News about our Work', both illustrated with male subjects drawings by the well-known Asturian artist Alfonso (Alfonso Iglesias López de Vivigo, 1910–88).[46] Verbal and visual paratexts guided the reader, providing a light-hearted touch to a publication 'whose writing style and contents might interest the largest number of readers',[47] although most actually came from industrial, scientific and technical sectors.

Accordingly, the formulas, tables, charts, notes and bibliographic references were interspersed with drawings illustrating highbrow studies, with male figures who were always labouring alone in their place of work, accompanied only by papers and instruments of science. Clearly, the epitome of the official scientific discourse was materialised here in the figure of the *vir modestus*, not only the stereotypical male image of the scientist *par excellence*, but also the public spectator, whose only function in the scene is to take in what is occurring or the results, but without interfering in any way (García Naharro, 2021).

Moreover, these drawings do not only illustrate, but also show what was understood to be the appropriate way to utilise a portion of the contents of the *Boletín*. This is embodied by the technical specifications section, topped by an illustration of a male figure, sitting alone at his desk, scissors in hand, only illuminated by a desk lamp, and cutting out the rectangular cards that materially comprise the physical pages of the section.[48] These 12 cm x 7.5 cm cards also included the title, author, place of publication and a brief description of the scientific articles, both national and international. In fact, these cards became such a significant part of the publication that the decision was made to publish them as a separate supplement beginning in January 1955. In that year, the *Boletín* issues contained a total of 484 printed pages, of which 144 corresponded to 'the technical specifications supplements, with 276 summaries of articles from a variety of national and international journals in the specialty'.[49] According to the institute's own reports, the journal was so well received that it soon increased the number of the periodical's pages. For instance, the six issues published in 1952 contained a total of 284 pages, which rose to 380 in 1953,[50] and then to 444 a year later. It was said of these pages that:

> they contain 268 cards for articles from different national and international journals in the field; seven original articles written by institute staff, seven written by other Spanish authors, two translations, 12 washability curves corresponding to carbon deposits produced by various other Asturian mines, 11 complete cards with the results of the assays and analyses conducted on samples of deposits from Spanish soft coal mines, five articles written to disseminate the results obtained from our work and a vast number of informative notes included in the 'News from our Institute' section.[51]

The proportion of authorship was almost always the same: of the eighteen articles included in the issues published in 1956 (issues 25 to 30), 'two were due to the collaboration of foreign specialists and five to Spanish authors without ties to the Institute', while the eleven remaining articles were written by institute staff, and presented 'diverse aspects of the work being done by the different Departments at the Centre'.[52] The same ratio characterised 1957, both in the number of summaries and articles, while in 1959, foreign specialists wrote four out of the total eleven articles[53] in the *Boletín*, clearly the natural route for publishing the institute's studies.

This printed product, then, consisted of a combination of different sections, with news and articles written by Spanish and institute specialists, and with translations of foreign works provided for the readers. They were also the recipients of the journal's translation and photocopy service – organised as a 'complement to the bibliographic Section of the *Boletín Informativo*'[54] – for articles included in the technical specifications section and pieces published in the journals sent to the institute; the different rates were sent to the interested party after the request was formalised.[55]

Finally, beginning with issue 14 (March 1954), the pages of the *Boletín* also made room for publicity related to the mining industry – Bandas transportadoras Pirelli, Máquinas y herramientas para minas ERIM, S. A., Aparatos y material para laboratorios Alonso Alegre y Cía and the like – that could be printed inside the journal (one page for 1,000 pesetas; half a page for

600 pesetas; quarter of a page for 400 pesetas), on the second and third pages (1,500 pesetas) and on the back cover (2,000 pesetas);[56] the last of these options was basically monopolised by advertisements for the German companies DKH (Didier-Kogag-Hinselmann, represented in Spain by E. Erhardt and Cía. Ltda) and Dr. C. Otto & Comp. (represented in Spain by the limited company OTTO). In practice, these were multi-purpose advertisements. On the one hand, while informing the reader about products or services of interest, they also paid for some of the journal's production costs. At the same time, the journal acted as an ideal showcase for the companies promoting themselves in the sector. They turned to the journal not just because – as with all official journals – the *Boletín* could count on the prime distribution channels and means provided by the regime, but also because of the symbolic and social capital of the institution behind the publication: the Spanish National Research Council.

Combustibles (Fuels): *Traces of the Benchmarks of the Franco Regime*

Another publication that relied on a vast amount of publicity was *Combustibles*, a journal published by the Institute of Fuels. Produced by the Madrid printshop DIANA Artes Gráficas (6 Larra Street) in a 21.5 cm x 27.5 cm format, the first issue appeared in Madrid in January 1941. The first editor-in-chief was the chemist Luis Bermejo Vida – a true defender of 'traditional, Catholic, authentic Spain, for which he smashed so many spears'[57] – who also headed the Institute of Fuels until his death on 19 September 1941, with Vicente Gómez Aranda acting as secretary. The writing staff and administration worked out of 2 Amaniel Street in Madrid, where requests for annual subscriptions were sent, priced at 30 pesetas for Spain, 35 pesetas for Portugal and Latin America and 40 pesetas for other foreign countries.[58] However, soon after Bermejo Vida's decease, the Institute of Fuels, which was originally located on the premises and in the laboratory of the Faculty of Science at the Universidad de Madrid, moved to the Faculty of Science in Zaragoza.[59]

As a journal that originated during a serious period of shortage in Spain, this publication endorsed the motto attributed to Franco linking the basic needs of the household (*'Ni un hogar sin lumbre, ni un español sin pan'*), putting fuel on the same level with the daily need for bread by adopting the oil lamp as its symbol, 'without which', the journal said, 'no modest private home can be'. Thus, from the very beginning, the silhouette of an oil lamp appeared on the covers of the journal, reflecting the various colours they used, much like the CSIC emblem on its back covers. However, although the first issue of this bi-monthly publication dedicated page one to a reproduction of image of the Caudillo as required, beginning with the second issue, the two first pages, two last pages and back cover were already reserved for publicity, paid for by companies in the sector. The prices for these advertisements varied in 1941 from 400 pesetas for a full page, to 250 pesetas for half a page and 150 pesetas for quarter of a page.[60] The advertisement pages flanked the contents of a journal that aspired to reflect 'everything related to the science, technology, economics and politics of fuels'.[61] To that end, the publication structured and divided its contents between original works, journal reviews – by topic: coal, fuel, gases and gas generators, etc. – and a book review section, along with 'an index of interesting articles . . . provided by the CAMPSA Bibliographic Information Service'.[62]

In 1942, after the journal moved its writers and administrative staff to the Faculty of Science at the Universidad de Zaragoza, it continued with the same format and structure, although the quality of its paper decreased when it began to be printed at the shops used by the newspaper *El Noticiero de Zaragoza* at 79 Coso Street. Under the new editor, Vicente Gómez Aranda, a chair professor at the Zaragoza Faculty of Science, the number of advertisements in the journal gradually decreased until they disappeared completely from its pages in mid-1943. However, this was not the only notable change over the years; although the journal since the beginning had planned to publish six annual issues, by 1942 it had been forced to reduce that number to three, albeit with 100 foliated pages each. For example, in 1946, issues 24–5, 26–7 and 28–9, corresponding to 1945, were published,[63] while beginning with issue 37 in 1947, only four issues of the journal appeared. In other words, *Combustibles* became a quarterly publication. At the same time, the Zaragoza section of the newly christened National Institute of Fuels also published the following monographic issues: *La síntesis Fischer-Tropsch. Estudio Bibliográfico* by Pascual Lahuerta Casaus (issue 21, eighty-five pages); *Carburantes naturales y artificiales* by Vicente Gómez Aranda (issue 22, forty-seven pages) and *Estudio sobre los carbones de la cuenca del Ebro. 1) Los lignitos de Mequinenza* and *2) Los lignitos de Utrillas* by Paulino Savirón Caravantes (issue 23, 280 pages).[64]

As Vicente Gómez Aranda himself observed, by 1949, the journal had published a total of forty-seven research studies and fifty-nine informative pieces, along with sixty references and translations, sixty-six reviews and 753 abstracts.[65] In 1950, this 28 cm x 22 cm publication, once again bi-monthly and containing nearly 500 pages a year, defined itself to its readers as a journal:

> that will keep you up to date with the advances in fuel science and technology around the world. *Combustibles* is the official publication of the National Institute of Fuels and presents the results of its own experiments. However, it also relies on the assistance and collaboration of eminent Spanish and foreign scientists and technicians, who add prestige to its pages. Its section of abstracts regularly provides news about the most notable publications from around the world in this specialty. *Combustibles* is of use to anyone interested in fuels: Industrialists, Financiers, Miners, Geologists, Chemists, Engineers, etc.[66]

Expert and technical readers, scientists and entrepreneurs in the sector were the focus of this publication, which modified its appearance in the early 1950s by changing the typeface of the cover titles. The word '*Combustibles*', now in thick black letters, was high up on the cover, while its association with the Juan de la Cierva Board – now 'of Technical Research' – was located on the bottom, leaving the centre of the page for the CSIC logo, a stylised *Arbor Scientiae* surrounded by the letters of the CSIC. In the meantime, as was becoming common practice, the classic institutional bookplate filled the middle of the back cover.[67] The journal temporarily lost its silhouette of an oil lamp, although it recovered it quickly for issue 55 (January–February 1951), which required moving the CSIC logo to the top, while the title returned to its place on the body of the cover.

Beginning in 1952, *Combustibles* also underwent some notable modifications regarding the type of paper and print type size, which was reduced, on its inside pages, as well as changing the fonts of the section titles: original works, courses and conferences; journal reviews; general information; abstracts; book reviews; bibliographic information and so forth. Some of the sections were illustrated using recurrent elements from the scientific-technical imaginary,[68] and advertisements were once again included in its pages. In 1952, the journal also published a general index corresponding to the first ten years of the publication, as well as a back issue for 1944, 'which was not published at the time, due to special circumstances'.[69] Moreover, while *Combustibles* had dedicated its first page to an image of the Caudillo ten years earlier, its issue 59 (January–February 1952) reserved its first inside pages to reproduce a translation by María del Carmen Gómez Parra from the Italian version of a speech given by Pope Pius XII on 22 November 1951 at the Pontificia Academia de Sciences, titled 'The Proofs for the Existence of God in the Light of Modern Natural Science'.[70] Clearly, despite the passage of time, the benchmarks of the Franco regime continued to be those of its official publications as well.

Finally, in January 1953 the spines of every issue began to include its name, the issue and its corresponding number, printed in black ink. The journal continue to be published without problems, putting out the corresponding number (62–8), maintaining exchanges of publications from the library of the Zaragoza section of the National Institute of Fuels and establishing a relationship with the Permanent Exhibition of International Scientific Books at Sveučilište u Zagrebu University in Yugoslavia.[71] Naturally, when the journal published issues 69–74 in 1954, the Zaragoza section made note of the foreign interest in *Combustibles*, 'demonstrated this year by the appearance of several original works by eminent foreign specialists on its pages'.[72] Ultimately, this was the function of these works in the corpus of the journal: the inclusion of foreign articles was meant to serve as a faithful reflection of the impact of *Combustibles* beyond Spain's borders. In reality, however, there is very little hard data in the sources to support this idea.

Ciencia y Técnica de la Soldadura (Welding Science and Technology): *Interactions of Social Uses and Material Forms by Usability Trials*

In 1947, the industrial engineer Francisco Fernández Aguado was the head of Publications Office at the Welding Institute, which while initially arranging a small library, at the same time agreed to a request made by an 'important English Publishing House' (Iliffe) to review 'the Spanish part of a dictionary in four languages – French, English, German and Spanish – dedicated to the vocabulary used in the field of welding and its techniques'. In Spain at this time, it was not unusual for many of the published translations of technical and educational works to first pass through the hands of the professionals at the CSIC for their supervision. In fact, years later, the Publications Office also assumed responsibility for the Spanish part of the *Multilingual Dictionary of Welding and its Techniques*, published in twelve languages by the International Welding Institute, copies of which could still be found in 1956. As advertised at the time, these copies were available, 'on a strict first come, first served basis', through their own channel, the journal *Ciencia y Técnica de la Soldadura*.[73]

In the late 1940s, the Welding Institute began to design a publications plan, developed through the publication of a series of thematic brochures, while at the same time setting up a Bibliographic Documentation Service 'for the members of the Institute and interested companies'.[74] By 1958, then, the Welding

Institute's agenda was consolidated and its editorial department could boast of a journal:

> published to date without interruption; we produce other monographic publications, we study and collaborate with other organizations preparing Standards and Specifications, we have prepared and published dictionaries in 12 languages, completing the Spanish section ourselves and 15 Committees work and collaborate with their colleagues at the International Institute of Welding.[75]

Ten years earlier, in 1948, the first issue of the welding newsletter, *Boletín de Información del Instituto de la Soldadura,* was distributed. This quarterly publication 'informed readers about what the Institute does and the tools for work and study that it offers', paying special attention to the holdings in the institute's library.[76] One year later, the same dynamic was at play, with '725 references with books and brochures and 534 summaries of many other journal articles',[77] and by 1950, this had risen to '811 book and brochure references and 1,100 article summaries'.[78] However, only one year later, in 1951, the first three issues of the institutes' bi-monthly journal, *Ciencia y Técnica de la Soldadura,* appeared, with an 'average print run' of 1,000 copies.[79] Produced in a 20 cm x 30 cm covered spiral ring notebook format with a metalised CSIC logo printed on the back, the new Welding Institute journal sold for 25 pesetas, with the annual subscription costing 130 pesetas, to be ordered from 58 Goya Street in Madrid, the institute headquarters. Together with a series of monographic publications, this journal would come to serve as the communication channel for the work being done by the institute's researchers, thus replacing the former *Boletín de Información.* The journal's stated goal was to:

> disseminate in our country the great possibilities of welding, combining its purely theoretical side with the practical side, in order to achieve effective results. Thanks to the classification of the subjects by material and a format especially designed for this purpose, it will be possible to organize bibliographic archives to facilitate the

> work of not only researchers and scholars, but of industrialists as well.[80]

In fact, the independent pagination by sections and the thick pages with four holes on the left side made it easy for – and even inspired – the reader to remove them for later archiving and classification. Furthermore, the institute itself encouraged this in printed notices that reminded its readers that they offered 'collections of filing folders to store the Journal articles for the price of 600 pesetas'.[81] Similarly, the pages in the technical specifications section, which contained eight 10 cm x 7 cm index cards, were pre-scored and ready to tear out, as clearly represented in the drawing of the welder that accompanied this section: a male operator who was also interspersed among the pages of the other sections – science and technology, welding jobs, standardisation, general information, news bulletin, patents, bibliography, work sheets – illustrating and giving the articles a more casual appearance.[82] However, these drawings soon disappeared from the pages.[83] The journal began to be printed on glossy paper with a finer finish and became less frivolous in 1954, when the Madrid printshop Langa y Cía. (6 Tahona de las Descalzas Street) took over the production of this publication, paradoxically designed to be taken apart.

Indeed, the format of the journal invited its readers to separate it into each of its parts, thus breaking away from the order imposed by the editorial composition to produce, if the reader so wished, a sort of 'anthology at the service of the user' (O'Donell, 1998: 48), in other words, compilations of diverse texts, anthologies of reading material that could constitute 'slapdash nonbooks' with 'texts containing different fonts, designs, reference conventions and even page orientations'. The journal thus took on a life of its own, materialised in a hypothetical new format, moulded by its users, which broke with the 'feeling of solidity and permanency' presupposed by the journal as device (Landow, 1998: 215).

However, during those years, not all the users were equal from the journal's perspective. In this respect, *Ciencia y Técnica de la Soldadura* had an initial print run of 1,000 copies – which rose to 1,200 in 1955 – that, the journal wrote in 1954, 'were sent at no cost to all the members of the Institute, official Bodies and Centres similar to ours', without being detrimental to (according to the institute, at least) 'some individual subscriptions in Spain and abroad'.[84] This publication sought to offer the 'technical and scientific precision that corresponds to a growing perceived need for national and international technological information and data to guide Spanish industry, while continuing to dedicate sufficient space to disseminating different welding methods, procedures and forms'.[85]

Industry also found a promotional space on the last pages of the journal (and beginning in the 1960s, on the first and inside pages, as well). Beginning with twelve pages in 1951, which featured publicity for Artac-Quijano electrodes, Philips, Amsa (Autogena Martínez S. A.), Fundor, Talleres Unión S. A. and the Sociedad Española de Carburos Metálicos, the number of pages filled by advertisements rose to sixty in the 1956 special issue that commemorated the ninth International Institute Assembly being held in Spain. Moreover, for the first time, this issue contained translations of its articles into French and English, in addition to advertising the Friden automatic calculator (represented in Spain by Rudy Meyer S. A.), the Madrid-based Talleres Kaeser S. A., the barrel company FEMSA (Fabricación de embases metálicos S. A.), SEAT (Sociedad Española de Automóviles de Turismo S. A.) automobiles, the Zaragoza-based enterprises Aparatos Gala and Talleres Jorda S. A., Crisoles Vulcano, General Eléctrica Española, Sociedad Ibérica de Montajes Eléctricos S. A. and three volumes of 'the most complete and modern *Tratado General de Soldadura*' by P. Schimpke and H. A. Horn, published by the Barcelona publishing house Gustavo Gili S. A.[86] Thus, the pages of this official scientific and technological publication also became a place to promote actors with links to the scientific field: specialist publishers, industries and companies that sought visibility in the pages of a journal that they saw as a circulating printed vehicle in the field of welding, both in Spain and abroad.

The journal plainly demonstrated its complex socio-material nature (Duguid, 1998: 88). At the same time that the advertisers, according to the journal, 'helped with their publicity to mitigate the costs inherent in a publication of this type',[87] the periodical also operated for them not only as a type of bulletin board, but also as a job pool of sorts. The advertisers could send the journal 'all the job offers and requests regarding technicians and workers who specialized in welding. The corresponding notices will be included in this Journal at no cost'.[88] The advertisements, published on loose papers or sheets, included jobs being offered or sought for workshop managers, industrial experts, industrial engineers, master welders, workers, specialised operators, sales agents, skilled workers, industrial draughtsmen and more.

Because of the companies involved and the jobs offered and sought, these advertisements and opportunities help to reconstruct the circuits through which this journal seems to have circulated and been visible. In fact, the institute was quite aware of the importance of the visibility, promotion and circulation of its product and that 'it needed to continuously "attack" and "bomb" the eyes and ears'. In 1958, the journal wrote that:

> Science is not money-minded, but it needs its propaganda at the same time. The dissemination of conferences, meetings, seminars, etc., both Spanish and foreign, must be as wide as possible. It must reach everywhere. Every location has a small industry, a little craft workshop, inherited from parents and grandparents . . . operated today by young brains, some unexposed to modernism, who will be awakened from their stagnation after reading or hearing about new technical implementations, studies or modifications.

The journal's readers vouched for this: 'Our Institute has received countless letters from these small workshops and industries requesting comprehensive information about how to improve their industry or endeavours.' In their letters, these readers thanked the Information Office – 'our best propaganda', according to the Welding Institute – for the help proffered. Moreover, as an official CSIC institute, this outreach work found essential allies in the national media: 'the Spanish are kept abreast of our activities thanks to the understanding and assistance we have unfailing encountered in the national Press and Radio.'[89]

For all that, *Ciencia y Técnica de la Soldadura* was a publication that, without compromising its promotion of all things national, made an effort to introduce foreign technology, as a guide to help develop Spanish potential. To that end, the journal offered summaries of specialist foreign publications, as well as 'technical journals and books from the other side of the Iron Curtain, that were procured by the Institute on a somewhat regular basis'.[90] As the journal itself emphasised in 1958, foreign scientific and technical ideas were vitally important, and an extra effort was made in that respect:

> Our readers will have observed that in the most recent issues of our Journal, we published twice as many technical specifications cards as we have usually inserted in previous copies. This is due, on the one hand, to the repeated requests made by many of our Members and readers to increase the number of technical specifications cards that we publish and, and the other, now that we are exchanging our journal with Russian journals with the same specialty, we have a larger number of very interesting articles, and we are constantly being asked for photocopies and translations.[91]

The institute was, however, well aware of the difficulties entailed in acquiring these publications in Franco's anti-communist Spain, 'due to the differences that separate us and communication difficulties'. At the same time, the journal recognised that:

> in the recent years since the War, the Soviet Union, like the United States and other countries, has made giant steps in industrial research and evidence of this includes, amongst other things, the artificial satellites that are currently flying at breakneck speed above our heads.

Hence, the journal made the firm commitment to follow:

> the example of the countries that are at the forefront of progress, and we know how to take advantage of the ideas of other peoples,

> without taking their political ideology into consideration, since material progress is common to all mankind, and their developments on behalf of the peace, well-being and progress of all are truly praiseworthy.[92]

This discourse of science as a 'neutral' activity was in complete harmony with the apolitical technocratic rhetoric of the era, when scientists worked (theoretically) beyond the dishonest and ambiguous motives and practices that characterised the field of politics (Nieto-Galan, 2019: 163–88).

Revista del Instituto del Hierro y del Acero (Iron and Steel Institute Journal): *Technical Specifications, International Cooperation and Boundary Work*

If 'researching and establishing the resources aimed at our industrial improvement have become the tutelary mission of the State', the creation of the Iron and Steel Institute corresponded to 'the mission to inform, project and implement industrial research work and studies'. Constituted at the time in the image and likeness of institutions such as 'the British "Iron and Steel Institute", the "British Iron and Steel Research Association", the "American Iron and Steel Institute", and the French "Institut de Recherches de la Sidérurgie"', this official Spanish institution – 'under the auspices of the Juan de la Cierva Board' of the CSIC – always enjoyed 'the support, in the form of opportune subsidies, for works . . . on steel and the iron mineral'.[93]

On 23 January 1947, the Iron and Steel Institute came into existence, under the leadership of the honorary colonel of the Armaments and Construction Corps of Engineers, Agustín Plana Sancho (1899–1990). Provisionally set up in some offices in Madrid, on the first floor of 13 Villanueva Street, the institute also had two separate sections, one in Barcelona and the other in Bilbao at, respectively, the Barcelona Institute for Technical Research and the Vizcaya Industrial Centre. Of the three departments that made up the institute's initial and hypothetical organisation chart, the first that was truly up and running was the Department of Technical Information (Crespo García *et al.*, 2009: 45–76). From there, priority was given to the library and technical specifications sections, trying to create, firstly, a bibliographic index for iron and steel, using references from various national libraries, while preparing another technical index with 'extracts from all the articles of interest on iron and steel that appear in the journals on hand'. At first, these index boards were for the internal use of the institute staff, but they were designed to comprise 'a section of our journal in future':[94] the quarterly publication of the Iron and Steel Institute whose first issue appeared in 1948.

Printed in a 21 cm x 29.5 cm format and with the CSIC emblem on its cover – a bas-relief *Arbor Scientiae* below the journal title – the first issue of the *Revista del Instituto del Hierro y del Acero* was printed by the Madrid company Artes Gráficas Langa y Cía. That issue was devoted to giving an account of the happenings at the Institute during its first year, making it 'formally and substantially different in content from what the publication was to be': a quarterly journal, 'with possibly future plans to increase its frequency'. It would include:

> whatever studies, reports, reviews and news that may be of use for our Members and Associates, to which end the contents will be duly arranged in five sections that will specifically include: technical cooperation articles, documents that summarize the Institute activities, technical information, industrial reports and the necessary bibliographic news.[95]

The journal also noted that the Department of Technical Information would be dedicated to gathering innovations related to iron and steel, both from its own institute and from others, also adding 'extracts and even full texts, when appropriate and possible, of works of renowned importance that have appeared in other specialist publications, thus making accessible to all our Members and Associates information that is usually out of the reach of individual initiatives'. This material would complement what, 'since the beginning of this year, the Institute has assembled under the name "Technical Specifications"',[96] an arm of the Department of Technical Information as well as its sections for "Translations", "Library and Publications", "Statistics" and "Foreign Relations"'.[97]

In 1948, the journal specified that the technical specifications cards, whose contents were mainly taken from the foreign journals received by the library and publications office, 'were made available to all our Members and Associates on a regular basis (monthly)'. Only a year after the institute was founded, there were already 539 cards of this type, 'a number that grows each month by some 60 references'.[98] This was clearly an important endeavour, and one that continued and expanded over time. In March 1952, the charge for photocopying the extracts of works in these cards was 10 pesetas, while microfilm ranged from 7 to 14 pesetas.[99] Moreover, a separate publication for these technical specification files, *Cuadernos de Fichas Técnicas*, was launched in 1948 for distribution amongst the institute members for 12 pesetas; by March 1952, it had published some 3,600 cards.

Table III.6. Total number of technical specification cards published to date

Year	**Total number of technical specification cards published to date**
1952	3,600
1953	4,559
1954	5,519
1955	6,319
1956	7,039
1957	7,759
1958	8,479
1959	9,199

Source: Juan de la Cierva Board for Technical Research, annual activity reports

Members, associates – 'whether groups (Corporations, Institutes, Companies, etc.) or individuals (technical or scientific Figures)'[100] – and industry were, therefore, the primary beneficiaries. They were also contacted by the institute to arrange print runs for copies, receive suggestions and opinions and, even, to complete the contents of the pages of the *Revista del Instituto del Hierro y del Acero*, 'especially the Collaboration and Industrial Report pages, so that whoever so wishes can make the results of their experiments, studies and research into subjects related to iron and steel available to others'.[101] This initiative was well received, and the technical collaboration articles were soon produced as offprints purchasable for the price of 10 pesetas in 1952 (as opposed to 30 pesetas for the regular issues and 50 pesetas for the special issues).[102] However, these 'members' were the 'only recipients' of the institute's publications, with the exception of 'out of catalogue' works, which could be bought by the general public.

Table III.7. Number of foreign organisations exchanging journals with the institute

Year	**Number of foreign organisations exchanging journals with the institute**
1951	20
1952	21
1953	40
1954	46

Source: Juan de la Cierva Board for Technical Research, annual activity reports

In addition to the photographs, charts, figures and small illustrations that accompanied the texts on its glossy pages, the *Revista del Instituto del Hierro y del Acero* also contained images loaded with the idea – seen throughout the CSIC's publications – of science as an activity separate from the common folk and confined to laboratories, far from the madding crowd. This solitary and individual activity was performed by the scientist – always represented as a man – with the sole company of his scientific instruments.[103] While this powerful image began to change after the Second World War in the countries taking the lead in research towards that of a group activity (Lafollete, 1990: 110–27), in Spain it prevailed[104] as a useful tool for creating an atmosphere of unintelligibility around science, as well as to reinforce the authority of the scientist in the official discourse.

Thus, the pages of the official journals also indirectly strengthened the stereotypical image of the scientist – a 'man', 'white', 'rational' and 'devoted' to research activity – presented at the time as obvious and natural. Concurrently, similar images continued to appear in other cultural products of the era, from cartoon figures – *Los grandes inventos del TBO* (Freixes and Garriga, 2008), the wise Megatón in the comic *Tío Vivo*, school textbooks like *Invenciones e inventores* (Solana, 1959) – to the images projected in the NO-DO newsreels (Ramírez Martínez, 2011). It is for this reason that working on the cultural construction of the image of science – and 'what is scientific' – is never a trivial matter. Rather, the study reveals how science was, and perhaps continues to be, an activity primarily defined as male, a constructed idea, in turn, built on strongly established prejudices transmitted through extremely diverse channels that carry sex and gender connotations that devalue women and extol men (Harding, 1997: 186–200). This range of images, moreover, is not only found in official scientific publications, but also in the words of the official discourse of the Franco period, thus reinforcing the idea of the traditional ethos of science and the *vir modestus*, with the consequent political and social implications.

In any case, regarding the composition of the *Revista del Instituto del Hierro y del Acero*, in 1949, the journal began to include separate sheets, such as a 'Fe-C Balance Diagram' and 'Sauveur-Type Table for Centigrade to Fahrenheit Conversion and Vice-Versa', documents unattached to the body of the journal that constituted, in the eyes of its editors, a 'collection of practical data that will always be useful'.[105] Each issue had one or two of these pages, printed on cardboard and containing data on hardness scales, annealing colour changes, diagrams, qualification tables and the like. These practical, collectable tables – which could be purchased separately for three pesetas[106] – at times came with two side perforations for easy archiving.[107]

Although the journal did not contain any publicity until 1960, its pages often included inserts advertising job openings or information about a variety of events. These ranged from an homage to the Reverend Priest Eduardo Vitoria Miralles (1864–1958), founder of the Sarriá Chemical Institute and 'his journal *Afinidad*, the most prominent among Spanish Chemistry publications',[108] to the International Conference on Smelting in 1955, the same year that Franco visited the institute's facilities and signed the visitors' book.[109] They also included the appearance of new publications, both products of the institute and those put out by technical publishing houses like Dossat, Labor, and Montecorvo, as well as those of interest to the field of metallurgy, like the English version of *STAL*, one of the most important Russian journals, and *Acta Metallurgica*, the international metal science journal launched by the American Society for Metals, which was co-sponsored by the Japanese Metal Institute (in 1953, the Spanish institute itself also began to cooperate with *Acta Metallurgica*, with works published 'in any language, and abstracts in French, German and English').[110]

However, this was not the only international collaboration on which the institute worked. For example, in April 1948, it made an agreement with the McGraw-Hill Book Company (New York) to translate and publish *Iron and Steel* by Hugh P. Tiemann,[111] with a view towards publishing it in 1949. With the translation work delayed, the book was finally published on 30 April 1950, a leather-bound octavo volume of 780 pages, written on scritta paper. Categorised as 'out of catalogue', it was sold for the price of 100 pesetas – 75 pesetas for institute members and associates – and the orders 'had to be handled on a rigorous first come, first served order, due to its limited number'.[112]

Iron and Steel was 'a small iron and steel encyclopaedia', a 'truly efficient English-Spanish dictionary of terms and technical words related to iron and steel'. It was expected to be warmly received, as was *Grundlagen der Eisengewinnung* by the engineer

Robert Durrer, a professor of metallurgy at the Eidgenössische Polytechnische Schule in Zürich and member of the Iron and Steel Institute. At the time, negotiations were still underway with the Swiss publishing house A. Francke A. C. regarding the Spanish-language translation and publication rights, which finally appeared in Spain in 1948 under the title *Fundamentos de la obtención del hierro*.[113] A few years later, in 1955, the second edition of *Métodos de Análisis Químicos Unificados* was announced, a 295-page paperback, offered at the price of 25 pesetas for members and 30 pesetas for the general public, who could order it from (or in) the official CSIC bookshop: the Medinaceli Scientific Bookshop.[114]

Other publications included classification and technological tables, distributed 'amongst the members and Associates of the Institute', the second part of *Métodos de Análisis Químicos Unificados*, which came out in 1951,[115] and five books published in 1952 and distributed in 1953.[116] Finally, in 1956, a publication entitled *Instituto del Hierro y del Acero: Lista de Miembros* 'was distributed amongst all the members',[117] although it was not until 1958 that the iron and steel colour chart, *Tabla de colores para el marcado de los productos siderúrgicos, noviembre 1957*, appeared, accompanied – completed and paratextually extended – by 'a wall mural on card stock, approximately 50 x 41 cm, suitable for placement in a visible location in any premises that may be interested'.[118] This publication, therefore, demanded the reader's involvement, not only with regard to reading, but to the practical potential offered by the deployment – both visual and material – of the wall mural in the (ideal) space of iron and steel industry edifices.

Revista del Instituto Nacional de Electrónica (Journal of the National Institute of Electronics): *Grappling with the Spread of Foreign Scientific Terms in Spanish*

It was at the end of 1957 that the first issue of the newsletter *Boletín de Información* was published by the National Institute of Electronics (INE), a quarterly publication with a 1,000 copy print run (upped to 1,800 copies, just one time, on the occasion of the Joint Meeting of the International Electrotechnical Commission in 1959).[119] The *Boletín* was a paperback publication, 21 cm x 27 cm, and printed by Gráficas Virgen de Loreto in Madrid. Its dark cover highlighted the red of the institute's emblem – the image an atomic nucleus with electrons orbiting around it – taking the place of the CSIC *Arbor Scientiae*, which appeared on the first page, along with the contents and the journal's institutional affiliations: the CSIC, the Juan de la Cierva Board, the National Institute of Electronics, the Board and the Editorial Board. These connections made this publication 'a jewel in the Institute', that is to say, an ornament that gave it lustre, assembled by redactors selected from among 'qualified volunteers who spoke languages and, until the establishment of the extended workday, devoted their Saturdays, which was a free day in the Laboratories, to this endeavour'. Moreover, they were paid using 'funds raised by subscriptions and advertising'.[120] In other words, all that glitters is not gold.

Although the name changed one year later to the *Revista del INE*,[121] the first issue of what was still known as the *Boletín* appeared in January 1958 at the price of 25 pesetas, defining itself as 'a documented and sober publication' that, nonetheless, aimed to be 'a respectable and agile publication', far from the 'avoidable stilted obscurity that makes us into a marble tower, inaccessible to most Spanish electronics technicians'. With the goal of 'enlightening', 'informing' and 'instructing' and – as they stressed even then, helping to prevent the exodus of Spanish researchers[122] – 'providing information about what is being done around the world and celebrating everything that can be done in Spain',[123] this publication opened its pages to the work of the institute. As the journal said, the function of this work was 'to create a national electronics technology and make it available to Industry and Services', thus helping to raise the visibility of the labour of the laboratories that, to paraphrase Santiago Ramón y Cajal – whose role in the first issue was quite prominent – would serve as the threshold to the factory.

For their part, national – and international – industries, companies and factories found a showcase open to the field of Spanish electronics in the journal's pages. Advertisements from companies like Iberavia S. A., Anglo-Española de Electricidad S. A., AGMA S. A., Pirelli, Cuarzo Radioeléctrico Español S. A. (CRESA), Vidrios Belgor S. A., Pla Hermanos y Cía. S. L., Marconi Española S. A., Laboratorios Bianchi, Standard Eléctrica S. A, Industrias Orbis, Compañía de productos eléctricos S. A. (COPRESA), Electrónicas Boar, Radiométrico and Electrónica Ibérica S. A. (ELIBE)[124] shared space with important dates, bibliographic cards, erudite citations – 'bouquets of select thoughts'[125] compiled by 'Fray Lito'– INE achievements and articles written by researchers at the institute and, at times, even the journal's own editors or writers from the CSIC in general. These authors were not only identified by their institutional affiliation, but also their curricular background, accompanied by a photograph that illustrated and – along with the article's bibliographic references, figures, equations and tables – gave textual form to their authority on the subject. These authorities were all male, faces and names of men who were always key leading figures on the pages of the journal (despite the occasional concession to the female sex, such as the effusive dedication to Luisa Arroyo Valdés, the disciple of the Professor Miguel Catalán, who in November 1962 obtained her doctorate with the highest honours).[126] In addition to prioritising the authority of the researchers at the institute and the council, the journal – renamed the *Revista Técnica del Instituto Nacional de Electrónica* in 1959 – also highlighted their corporative nature, not only by providing information about the institute's accomplishments and including publicity inserts from the Juan de la Cierva Board, but also by showcasing its documentary endeavours by including reviews of the works done and compiled in the institute's technical reports.

The INE researchers were required to deliver these reports, along with the assistants' measurement logbooks, on a regular basis, as a way 'of leaving a written record of the studies and developments that had produced positive results'. The lab work reports were to be written following clear norms: 'A brief, explanatory title, a cursory table of contents; an introduction with a general presentation of the problem and the state of the art before beginning the study; the theoretical foundations and the method used to tackle it; conclusions and the consulted bibliography'. Once drafted, six copies were ordered for distribution to the head office, the library and the four research groups, leaving them 'freely available to the entire INE, and when the head office deems one of them to be of sufficient interest, the author will be invited to lead a colloquium on it with the INE research staff'.[127] However, the first six technical reports were not published until 1959, including one by the aforementioned female researcher Luisa Arroyo Valdés on fluorescent screens: *Pantallas fluorescentes para tubos de rayos catódicos. I. Tecnología y resultados generales.*[128]

Nonetheless, in the words of the editor-in-chief of the journal, the National Institute of Electronics needed 'a voice to present it, to praise it and, if necessary, defend it. The Research Centres that have nothing to say and let years ago by without making a name for themselves will inevitably die forever'.[129] He went on:

> To think that a Journal can simply consist of a collection of articles, however valuable they may be, lined up one after the other, is a crass error. The Journal must have a heart, and having a heart means having personality; in other words, aspiring to a definite aim and guiding all its pages towards that aim.[130]

From this initial conception, the idea emerged that the *Revista INE* should achieve some so-called doctrinal goals for scientists and specialists, in short, an institutional affiliation. To that end, the journal – which in the early 1960s had 'already managed to acclimate to the scientific and technical circles of Spain and Latin America'[131] – published a special issue to mark the twenty-fifth anniversary of the CSIC: *25 Años de Paz. Aniversario del CSIC.* Thus, in October 1964, the journal – under the direction of José R. de Gopegui Gil, with a print run of 5,600 copies and four annual consolidated issues, an annual subscription rate of 100 pesetas in Spain and 175 pesetas abroad, and with an authorised branch in the Carlton Berry Company (Grand Building, Trafalgar Square, London) – published a 300-page volume celebrating the silver anniversary of the council and providing an account of all the work done by the institute. Of course, this was all done without abandoning the traditional sections; the eighty pages dedicated exclusively to electronic terminology vocabulary and including a compilation of everything published to date were particularly outstanding.

In this regard, by 1962, when the journal turned five, it had already produced a list of 'more than 300 cards with uncertain terms with proposals that, for the most part, have been accepted and incorporated into the language.[132] Moreover, the way to use this 'incipient electronic lexicon' was also explained in its pages:

> On the left side of the page, the original term is written in small capital letters, with a symbol whose first number indicates the Journal in which the corresponding card was published, and the second, that which preceded it . . . The definition of the concept is in the middle of the page, expressed or represented by the corresponding term or word. As it is to be part of a lexicon, this definition is concise, comprehensible to the vast majority of readers and is not necessarily rigorous, although it is free of errors, up to the point reached by our analysis of the question . . . Finally, on the right side of the page, and also in small capital letters, is the word or words that we deem most suitable to be expressed in our language, whether the concept or the defined meaning of the original word.[133]

To a large extent, that was the sought-after aim: 'To define the concept, suggest the word or term that we deem suitable to refer to it in Spanish and to spread it amongst the readers of the INE who, evidently, comprise the most numerous cluster of individuals who are going to need it.'[134] What was at stake, then, was looking after technical language and avoiding 'the mystification of the vernacular language with a progressive and growing invasion of foreign terms'.[135] In this respect, and following one of the suggestions that Esteban Terradas made in his acceptance speech before the Academy of Language, the journal dedicated its section titled 'Terminology' to firmly grappling with 'the problem of neologisms, publishing cards that define the concepts, propose the correct words to express them and open them up to public debate'.[136]

And, indeed, a public debate was soon underway in the journal's pages in the form of letters sent in by the INE's readers to the editorial office at 54 Ríos Rosas Street in Madrid. Letters came, for example, from the distinguished physicist Julio Palacios, who proposed that the word 'spin' be modified and written as 'ESPIN, which can be used to form the plural form ESPINES'; card No. 4 was thus rectified in the first issue of that section.[137] Although some writers preferred not to reveal their identities, the names of the known participants in the successive debates in this section include, in addition to Palacios, the engineer and advisor to the National Institute of Electronics, Antonio Colino López, Morgenroth Orozco, the meteorological engineer at the National Centre for Telecommunications Studies in Paris, a *monsieur* Pierre Misme, the electronic engineer from Submarine Cable Ltd, Mr K. E. Latimer, artillery captain José María Suanzes Siljestrom, the technical director and engineer at Transradio Española, Luis Cáceres, Professor José María Pardo from the School of Telecommunications Engineers, who was also the head engineer at the Spanish Frequency Registration Board and secretary of the National Telecommunications Board, and the journal's editors and writers, who offered their own proposals in debates that appeared in publications like *Radiovisión, Revista de Telecomunicación, Las Ciencias, Electronics* (United States), *VDI Nachrichten, Revista General de Marina, ABC* and *Blanco y Negro.*

These readers – all men – who spoke up in this arena were invested with the authority conferred on them by their professional profiles and institutional affiliations. Naturally, along with these factors, others, like gender and expertise, entered into play when it came to defining who had an authorised and legitimised voice in the field. At the same time, both the profile of the readers and authors of the journal and the dialogue held with other publications, as well as the list of advertisers from the field of electronics, provide an important set of keys when reconstructing, in practice, the inscribed circulation reflected in its pages. These include spaces, places, institutions, companies and editorial products that, together with the information provided by the empirical readers – socio-professional profiles, opinions proffered, debates and so forth – trace the itineraries in the text along which the journal circulated and, ostensibly, would circulate.

Finally, within the framework of the dictatorial regime whose official discourse advocated for a specifically Spanish science in a context characterised by suspicions about the uncontrolled eruption of foreign terminology in the language of science (Garriga, 2019: 145–61), in 1965 the journal's 'Terminology' section received the praise of the Royal Academy of Natural, Physical and Exact Sciences, whose own journal had also begun to publish a section on vocabulary.[138] Indeed, the 'Terminology' section had its counterparts in sections like 'Scientific Vocabulary' in the journal on nuclear energy, *Energía Nuclear,* which 'found its inspiration in the same postulates that informed ours', wrote the INE, congratulating itself for the initiative. For, according to the institute, 'without the indefatigable assistance of technicians', since the mid-1960s, 'it would not be possible to contain the deluge of foreign words invading our language'.[139]

Revista de Plásticos (Journal of Plastics): *Overseeing Scientific Works and Translations Produced in the Spanish Publishing Field*

Directed in 1964 by Juan Luis de la Infiesta Molero, the Department of Plastics, an arm of the Juan de la Cierva Board, dated back to October 1947, when the plastics section was created as part of the Alonso Barba Institute for Chemistry. Renamed the Department of Plastics and Rubber Technology in 1956, this department was 'oriented towards training staff, providing information and developing research in the field of polymers'.[140] Even though it did not originally have its own building, instead occupying offices in the Institute of Chemistry and the Torres Quevedo Institute, it did have its own journal, 'the only Spanish publication on this specialty and highly successful in the industry'.[141]

This publication was the *Revista de Plásticos,* which began to be published by the department in 1950 as a bi-monthly journal with an annual special issue. Its aim was to serve as a nexus with industry and provide the institute with some financial relief, not so much from its sales as the exchanges with (always expensive) foreign scientific journals, removing the need 'for our library to pay for any subscriptions, since its needs are covered by this exchange'.[142] On its brand new red cover, the journal's title was written in stylised black letters and, at first glance, its institutional ties could only be detected in the dark silhouette of the Iberian Peninsula, from which sprang the CSIC *Arbor Scientiae* in white. This first issue, published in January 1950, went on sale for 15 pesetas, with an annual subscription costing 80 pesetas. Printed by C. Bermejo in a 22 cm x 30 cm format, under the plastics section of the Juan de la Cierva Board for Technical Research, the journal expressed 'its most committed greetings in this first issue to the Caudillo of Spain, a token of the allegiance and affection towards one to whom Science and Spanish Technology owe so much'. These words appeared in its very first editorial, in which – also aware of the need to 'familiarize the public, using a type of propaganda, about scientific accomplishments and the spirit and character of science' – the journal offered to act as a showcase for industrial companies, from whom it also requested understanding, collaboration and encouragement. In February 1951, Juan Antonio Suanzes told the Plenary Session of the CSIC that:[143]

> The contact between this Section and industry is very tight, and has enormously benefitted the special nature of this journal, which simultaneously caters to science and technology without delay, with a design that is right and proper, moreover, to the newness of this type of industry. Many are the cases of advice and consultation handled by the Section, and the number of microfilms and photocopies supplied by its special library totals nearly 2,000.[144]

Beginning with the very first issue of the journal, these industrial firms advertised on its pages, filling both the periodical's back cover and its first and last pages, and in some cases, like the advertisement for the Gipuzkoa factory Resinas Sintéticas S. A., even appearing on the journal's contents page. Publicity for metal stamps (Fornituras Mabel), chemical pastes for the fabrication of plastic materials (Celupal S. A.), plastic moulding presses (Asimer S. A.), rubber moulding presses (Caucho Especial M. R. M, S. A.) and machines for plastic industries (Distribuidora DIAPAM) all shared space with publicity for industrial firms, such as Unión Química del Norte de España S. A., Plásticos Españoles S. L. and Central Española de Aplicaciones Científicas (CEDAC), as well as the latest publications of interest, like the Spanish version of Dr Roelof Houwink's work on plastics technology, *Tecnología de los Plásticos*, a fabric-bound, 18 cm x 26 cm volume priced at 1,000 pesetas, exclusively distributed to booksellers by the Dossat publishing house.[145]

Another recent publication was the work on modern plastics by Harry Barron, whose Spanish translation of the second English edition was done by Luis Graupera Castañé and published in 1952 by the Gustavo Gili publishing house.[146] The pages of this work also included an advertisement for the *Revista de Plásticos*, in compliance with a stipulation in the agreement with the publisher; in exchange, an advertisement and review of the work were to be published in the CSIC journal. However, as the journal confidentially informed the publisher, 'the review was going to be somewhat harsh and critical, especially regarding the translation'. In this respect, the Gustavo Gili publishing house observed:[147]

> It appears that the Plastics section of the Spanish National Research Council intends and is ensuring that all the works on that topic pass through its hands, and that they wish to provide and assign the translators themselves, in other words, they want to be given oversight over all the works that are published. I do not know if, in the present case, this had some influence when they judged the translation of the work in question, but it very may well could have.[148]

'There is an element of that . . . and it could be that the "beating" they gave the work was in some way based on that', agreed Joaquín de Oteyza, who had already unsuccessfully tried to convince Juan Luis de la Infiesta – 'the real top boss at this Journal' – to halt publication of the appraisal: 'Nothing can be done now.'[149] The review of the book, which was conducted by the secretary of the CSIC Department of Plastics, J. Fontán, emphasised its basic errors, about which, he said, 'saying nothing would endanger the trust between the Journal and its readers'.[150] This opinion was shared by the chemical sciences doctor Salvador Terol Alonso, who told Pedro Reverté – the head of the Reverté publishing house and former collaborator with the Gustavo Gili house, who had also tried mediate in the matter – that 'the editorial office of the *Revista* "Plásticos" has more than enough reason to express itself in such terms about a book that, while excellent in the original text, has a translation that leaves much to be desired'. He continued:

> While Fontán was showing me, one after another, technical confusions, unforgivable mistakes and elementary errors in chemical and technological terminology (strictly within the domain of the translation), I lost all my good wishes to insist on reasons of friendship and, even, our mutual interests, with regard to rectifying his notes.[151]

J. Fontán wrote back: 'Given the nature of the help and guidance that this Journal offers Spanish industry, we are obliged, although it pains us, not to keep silent about what could be a source of errors.'[152] Incidents like this one demonstrate the symbolic value of the *Revista de Plásticos* – and the opinions that appeared in it – for its advertisers. At the same time, it was clearly important for a scientific and technical commercial publishing house to maintain good relations with those in official circles. For that reason, despite everything, the Gustavo Gili publishing house did not hesitate to recommend the journal to the president of the Buenos Aires-based Meryland S. R. L. company when it requested information in December of that same year about a journal on plastics 'with propaganda from specialist manufacturers in this field, in order to investigate the possibility of subscribing'.[153]

Thus, advertisements and reviews of technical works shared pride of place in the pages of the *Revista de Plásticos*, alongside not only innovations in plastic and 'all the photographs sent to us of articles made of plastic that exemplify our industry', but also the articles – illustrated with images, tables, figures and bibliographic references – produced by the staff of the plastics section and consultations from the journal's readers who, beginning with the second issue, could send in 'any questions related to any topic of interest to the industry, whether scientific, technical or financial'.[154] This clearly demonstrated that industrialists and technicians were the ideal target audience for this journal, where they could also present their own bids and requests at no cost: 'in order to facilitate the exchange of products and machinery related to the plastics and / or rubber industry, we will publish all the notices we receive'.[155]

All these sections, combined with financial information, the newsletter, the bibliographic section (book reviews) and the technical documentation (reviews of specialist articles) shaped a journal that, in 1957, was being distributed not only by its own editorial team located at the Alonso Barba Institute in Madrid, but also through its delegation in Barcelona, under Luis Puig Peiró; in Portugal, through Portugalia Editora; in Argentina, through Librería Editorial Suc. F. Giménez Codes; and in Mexico, through Reverté. All these places, as well as 'all the Chambers of Commerce, embassies, foreign associations and journals in our specialty', were also sent the 1957 yearbook, *Anuario de Plásticos,* a work presented as 'the best example of the great developments attained by the Spanish Plastics Industry', editorially enshrined in the creation of an annual directory of that industry. This text included everything ranging from the names and addresses of the companies in the sector to the representatives, manufacturers and importers of the raw material, machinery and equipment used in the plastics industry, as well as a compendium of the foreign firms represented in Spain and, of course, the names of the primary actors in the spanish plastics industry. This onomastic index would circulate – along with information about agencies like the Spanish Plastic Centre and the Department of Plastics itself – through privileged channels of communication and amongst the national and international recipients of these yearbooks, which never ceased to be editorial products designed to promote and provide information about the sector, financed, in part, by the industries that advertised on their pages.[156]

Grasas y Aceites (Fats and Oils): *Visibility and Propaganda in Exchanges with Foreign Readers*

Published in a 16 cm x 22 cm format by the Special Institute for Fats and their Derivatives (8 Santa Clara Street, Seville), the first instalment of the quarterly journal *Grasas y Aceites. Información bibliográfica sobre su química, tecnología y aplicaciones* saw the light of day in March 1950. The institute's first periodical was dedicated to:

> providing the most complete bibliographic information possible on the global progress related to the chemistry and technology of fats, introducing all the Spanish and foreign research that may be of interest to our industrialists, due to the nature of their work and their trade relations abroad.[157]

As a publication 'meant to provide a permanent link with industry and with other research agencies, and to represent the work being done at the institute',[158] industrialists and researchers were its target readers and the chosen recipients of the institute's specialist library. As the journal itself noted, this was clearly 'an indispensable tool for the development of the work of our researchers', also put at the disposal of 'industrialists and scholars, so that they can read up on the subjects that are missing information in their own Centres and working laboratories'.[159]

Although the provision of 'bibliographic, analytical and research services by contract to private industry and individuals' was not established and regulated until 1954,[160] from the very beginning, the journal acted as a link between industry and the laboratories, containing 'the most important works published in the Journals received by this Institute, which are considered based on the following': 'The extracts are carefully written by specialists, with sufficient breadth for perfect comprehension, reproducing the necessary graphs and tables, with the authorization of the editor of the original publication.' The journal went on to say that works considered of secondary interest were shortened, to give the reader an idea of the paper or note they discuss.[161]

From some forty periodicals that the journal originally received, it extracted pieces of various lengths and grouped them into: 'Fats in General', 'Treatments', 'Fatty Acids', 'Drying Oils', 'Vegetal Fats', 'Animal Fats', 'Edible Fats', 'Detergents', 'Emulsions', 'Soaps', 'Antioxidants', 'Paints' and 'Splitting Reagents'. The journal also incorporated reviews of international conferences of interest, such as the Paris meeting of the International Commission for the Study of Fats in September 1950 (vol. I, fasc. 3) and the German Society for Fat Science conference held in Munich in October of that same year (vol. I, fasc. 4).

In 1951, and under the leadership of its editorial board – Juan M. Martínez Moreno, José María Rodríguez de la Borbolla, A. Risueño, Jaime Gracián and A. Vioque – the journal was modified in form and substance. Printed on glossy paper, in a 21 cm x 27 cm format, it sold for an annual subscription price of 50 pesetas in Spain, 60 pesetas in Latin America and 80 pesetas for the rest of the world, while incorporating references to national and international books on fats, a list of patents registered in Spain and abroad, as well as original articles and research works by the institute staff, as described:

> It must be of significant interest for many of our readers, because the bibliographic extracts, whether long or short, that we have published to date and that, naturally, we want to continue to offer, often require in-depth knowledge of the topic if they are to be of any use. This knowledge of the most important subjects in Fat Technology and Chemistry is what we will offer in a range of articles that we will try to write as clearly and accessibly as possible, even for non-technicians.[162]

By 1952, single issues of the journal were being sold for 30 pesetas in Spain, 35 pesetas in Latin America and 40 pesetas in the other countries, while annual subscription rates had risen to 80, 100 and

125 pesetas, respectively. The prices were the same one year later, when the journal opened its first fascicle with an image paying homage to 'His Excellency the Head of State', who had inaugurated the new institute building on 23 April. An illustration of the building graced the cover of the journal, whose 'specially bound' issues were gifted to the Caudillo, along with the institute's other publications.[163] After the head of state signed the visitors' book – as did all the ministers present at the act – the first fascicle of the year was presented to the attendees, a special number (vol. 4, fasc. 1) that described work done by the institute during the six years since its creation. Among other data, this included an overview of everything published by the institute staff, from technical reports printed in the journal *Anales de la Real Sociedad Española de Física y Química* (1948) to brochures 'distributed to industrialists and anyone else interested in our activities' (1949) and, finally, the appearance of *Grasas y Aceites* from its original format to the 1951 modification, 'in a new 22 x 28 cm format and printed on luxurious coated couché paper, with an abundance of drawings and photographs'.[164]

It was not until 1954 that the journal began to include advertising on its sober pages, specifically on the back of the contents page and on the last page of its issues. For instance, that same year, *Grasas y Aceites* included advertisements for the Seville company EXISA (Exportaciones e Importaciones S. A.), CECA (Compañía Española de Carbones Activos S. A.) and the suppliers of laboratory and scientific material, David y Torres S. L. in Barcelona and Casa Torrecilla in Madrid. Additionally, the fortnightly journal on olive growing, *Almazara. Revista Española de Oleicultura y de las Industrias Oleicolas*, published in Madrid (24 Sagasta Street, 6), was advertised on the journal's pages on a very regular basis; its publicity made it clear that, unlike *Grasas y Aceites*, it did not represent 'any entity or agency'.[165] As publications like *Almazara* and *Boletín de Oleicultura* were, in the words *Grasas y Aceites*, 'in more direct contact with economic problems than we are', they 'contained many Official Directives, replies and counter-replies and comments for all tastes', especially during the season of the olive harvest.

However, concerns about the olive harvest were also felt in the Institute for Fats and Oils beginning in late 1954, right when it began the project 'to operate its own experimental oil press as a testbed for inventors and machinery owners'. As the journal itself stressed, 'the concerns affecting the Institute at this time are not all purely scientific':

> This is the time, so important and feared, to draw up budgets; to take stock of what one has and prepare ambitious work plans on the basis of what is wanted . . . It is the time to present new needs and carefully reconsider research plans in order to deduce their minimum economic requirements. But administrative processes are always slow, and amidst this uncertainty, from setting out what is needed until knowing how to address those needs, several months will pass, and they are, precisely, the most active months of the harvest season, during which we are operating blindly, without knowing, in the end, what awaits us: prosperity or a serious economic imbalance.[166]

While the motivation behind including advertising in the journal was financial, it also clearly allowed it to increase the number of both its pages and its issues, from fourt to six a year.[167] The same economic motivations led the journal to encourage publishers to send copies of their works – at no cost – in order to expand the library of the Institute for Fats and Oils. In exchange, the journal offered to 'publish a lengthy review' of the texts, always done by 'a specialist on the subject'.[168] In other words, once again, visibility and propaganda were provided in exchange for new books, an advertising business that benefitted both parties. Advertisements, advertisers and the pages dedicated to them multiplied over time, reaching fifteen pages in 1958, with advertisements from local and regional industries, such as Ángel Camacho Alarcón, an exporter-importer of oils, soaps and olives, Morón de la Frontera (Seville), Hijos de Ybarra S. A. de Sevilla and Carbonell y Cía. de Córdoba. However, the geographic representation was even more extensive, taking in the Santiago Domenech mechanical construction workshop (Barcelona); Capillery oil filters, made in Spain under the 'Capillerys' patent in Le Vigan (France) by Capillery Española (Reus, Cataluña); Hans T. Möller, S. A. centrifuges (Barcelona); citric acid 'against the hardening of soy, olive and cotton oils', manufactured by EBRO Compañía de Azúcares y Alcoholes S. A. (Madrid); auxiliary products for the industry, such as PULCRA Ltda. (Barcelona), and international companies like Meccaniche Moderne (Italy) and The VD Anderson Company (United States). All these enterprises hoped to find future clients and users for their products among the journal's potential readers. Machinery, technology and services filled most of the advertising pages in the journal, which also served as a showcase for other scientific publications in the field, including the Italian *Olearia. Rivista delle Materie Grasse*, the French *Revue française des corps gras*, the official publication of the Institute for Fats in that country, and the Belgian scientific and technical journal, *Revue des fermentations et des industries alimentaires*, which charged 400 Belgian francs for foreign subscriptions.[169]

These companies and publications – both Spanish and foreign – believed that their presence on the pages of *Grasas y Aceites* would do more than attract Spanish clients. The journal was distributed and exchanged with foreign readers in mind, as demonstrated by the fact that, beginning in 1956, it included abstracts of its articles published in Spanish, German, French and English.[170] Indeed, and according to the institute's own data for that year, the journal engaged in exchanges with more than eighty Spanish and international publications. One year later, in 1957, the journals that were exchanged with *Grasas y Aceites* included *Alchimist* (Belgium), *Anales del Instituto Municipal de Higiene de Barcelona* (Spain), *Bolletino Scientifico della Facolta di Chimica Industriale* (Italy), *Bulletin Africain des Matières Grasses* (Marruecos), *Bulletin Agricole du Congo Belge* (Belgium), *Conserve e Derivati Agrumari* (Italy), *Elaboraciones y envases* (United States), *Industria Conserve* (Italy), *Industrie Chimique Belge* (Belgium), *Información Conservera* (Spain), *Laboratorio Oceanográfico* (Spain), *La Chimie des Peintures* (Belgium), *Oil and Soap Egypt* (Egypt), *Przemysl Spoczywczy* (Poland), *Recherches* (France) and *Revista de Plásticos* (Spain).[171]

In fact, the International Society for Fat Science designated *Grasas y Aceites* their Spanish periodical, and in 1960, it was exchanging issues with 130 journals – 'an updated list of which appears in the first fascicle of each year' – in thirty-nine countries and could take pride in appearing, since its earliest days, in the journal *Chemical Abstracts* (1907) of the American Chemical Society, the most important international publication in the field (which would dethrone the hitherto hegemonic, pioneering German journal *Chemisches Zentralblatt* (1830)).[172] At the end of 1960, *Grasas y Aceites* appeared as one of 600 journals listed in the new publication of the American Chemical Society, *Chemical Titles*, 'an objective demonstration', according to the Spanish journal, 'of the significant international standing reached by Spanish research into fats'.[173] Once again, this demonstrates the importance of taking journals seriously as objects that are exchanged and circulate between different social spaces, thus creating alliances and connections between a variety of actors both within and outside the Spanish scientific field under Franco.

Chapter 12

Coda – The Hidden Side: Foreign Perceptions of Science in Franco's Spain

In 1949, the journal *Science* published an article by F. S. Boig, 'Domestic and Foreign Periodicals in the Field of Organic Chemistry: A Statistical Analysis', which drew on data from the journal *Chemical Abstracts* to assess the global situation of the specific field of organic chemistry after the Second World War.[1] One of the most conspicuous conclusions of this article was the preponderance of English as a scientific language, as well as an increased consideration of other languages like French, Finnish and Spanish in the field. One year later, Manuel Lora-Tamayo proudly referred to this acknowledgement in his speech on research in organic chemistry during the closing session of the eleventh plenary meeting of the CSIC:

> This progress in Spanish research in the field of Organic Chemistry has been included, to our satisfaction, in an article in the American journal *Science* containing a statistical analysis of the publications coming out of various countries in this specialty. Using the abstracts published in *Chemical Abstracts* as the base, only 15 countries were selected from the whole world, and their publications and their percentages in 1947 were tabulated and compared to 10 years before. Spain comes in twelfth place, immediately behind Sweden and ahead of Finland, Canada and Austria, *having risen 10 places* in the last 10 years. The number of Organic Chemistry articles published in the Spanish language rose from 17 to 48, and the journal added that . . . 'the increase in the number of abstracts in the Spanish language is due to an increase in activity in Spain itself, rather than in

> Spanish-speaking countries', which is, moreover, clearly seen in the collation of published tables.[2]

However, writing in the same American journal nearly a decade later, US scientist David M. Gates (1921–2016) presented a very different perspective, closer to the image of Spain as a scientifically and economically backwards country under the yoke of an ultra-Catholic dictatorship. This image fully conformed to what was being presented in the United States, dating back at least to the late 1940s.

> 'Unfortunately, the situation tends to perpetuate itself.' With this phrase, the American physicist D. M. Gates ended one of his most interesting paragraphs on his view of the scientific situation in Spain – which he considers quite gloomy – published in the journal *Science* in August 1958.[3]

Thus began an editorial in the Spanish scientific journal *Grasas y Aceites*, in which Agustín Vioque repeated the assessments made by Gates, a physicist at the National Bureau of Standards' Central Radio Propagation Laboratory, and previously the scientific director of the London branch of the Office of Naval Research. Gates described a panorama of elegant laboratories lacking results, where the research funds were controlled by the Spanish National Research Council under the auspices of the Ministry of Education. The laboratories, which allegedly worked in alliance with institutes and boards that bore the names of famous Spanish scientists, in practice operated, he wrote, as independent laboratories. The endemic poverty in the country, the low salaries, which forced some professionals to take on a second job, and the lack of sufficiently prepared research staff were aggravated by the need at the time to import supplies and equipment, as well as the brain drain as scientists sought better working conditions outside Spain.[4]

In view of this situation, Vioque recognised in his editorial that the lack of results 'was a reality 10 or 15 years ago, and the fruits of the later efforts made by the new research institutions cannot yet be judged', but asked at the same time, 'Why have there been so few results?' Vioque did, however, then venture to identify some details, 'especially in the case of applied research', which could explain some of the problems endemic to national research. For example, he mentioned an 'absence of organization and, especially, coordination, cooperation', as well as the lack of private financing on the part of certain industrialists who preferred to look abroad before backing someone 'from home'.[5] This diagnosis coincided with that made by the Scientific and Technical Research Advisory Committee (*Comisión Asesora de Investigación Científica y Técnica*, CAICYT), according to which the lack of 'a coherent scientific policy' was exacerbated by the absence of a 'research spirit' in Spanish industry, which always found it more profitable to import foreign technology (Delgado Gómez-Escalonilla and López García, 2019: 18). To quote an editorial by the Spanish chemist Jaime Gracián, also written for the journal *Grasas y Aceites* in 1960:

> If this is not a question of research being denied or its advantages ignored, the only possible conclusion, thinking logically, is that we are dealing with scepticism towards Spanish science or, in other words, Spanish scientists. This also explains the favourable reception given to everything related to international science, and the satisfaction with which technology from other countries is procured and paid for, while the work of Spanish graduates is underrated and underpaid.[6]

The article by Gates attracted other responses, including one from Antonio García Verduch (1924–2009) of the CSIC Institute for Edaphology and Plant Physiology, whose answer was published in the *Letters* section of Science in July 1959. Verduch emphasised that the field of agricultural research did, indeed, receive funding and support from the CSIC, and invited Gates to read the 1954 booklet, *Actividades Internacionales del Consejo Superior de Investigaciones Científicas*, where he would discover how many CSIC researchers had research experience abroad.[7] Again, fellowship exchanges, invitations from and to foreign professors, attendance at international conferences and the exchange of publications with foreign research centres were then used as proof of the council's good work, as had been the case since the early 1940s, when the Edaphology Institute under Albareda had operated as a *de facto* improvised foreign affairs office (Nieto-Galan, 2019: 155–62).

However, this competitive and apolitical view of research under Franco also attracted responses from two researchers in the History Institute at Ciudad Universitaria de México, Juan Comas and Santiago Genovés, who used the same high-prestige journal in October 1959 to refute the representation of a free exchange of ideas and knowledge in Franco's Spain. These researchers reminded the readers of *Science* of both the strict censorship system that continued in place in Spain and the difficulties faced by scientists when they wanted to accept certain theories that were widespread in the Western world. They specifically mentioned evolution, presenting as evidence the anti-evolutionist claims defended in the first volume of *Historia Universal* (Espasa-Calpe, 1954) and the prologue to the work by Valeriano Andérez Alonso on the origin of man, *Hacia el Origen del Hombre*, which was written by Bermudo Meléndez, a professor of palaeontology at the Universidad de Madrid and director of the Department of Palaeontology at the National Natural Science Museum.[8]

Clashing viewpoints, contradictions and tensions once again demonstrate that reality is never so homogenous nor history so simple. During the 1950s, for example, while the council's Franciscan physicist, Friar Juan Zarco de Gea, talked to the Spanish press about the advantages and dangers of the 'peaceful use of atomic energy',[9] citing the maxim of science 'as an emanation of God';[10] almost at the same time, in the inner circles of the second Poblet Conversations of the Menéndez Pelayo Association, the degree of authority that the Catholic Church should have in terms of evolutionism was being challenged. However, these doubts were never expressed outside the elite circles of this association or appeared in the press or specialist journals of the time (Florensa, 2017: 411–12).

Neither did the real reason that José Ibáñez Martín published his October 1961 article on 'Franco and the Spanish National

Research Council' come to light, which was simply to respond 'to most of the objections[11] made about the Council, although not yet to all of them'. Writing in a letter to the technical secretary-general of the Presidency of the Government, Laureano López Rodó, Ibáñez Martín added, 'I hope later to fully answer those that are characterized by their injustice'.[12] When this piece ended up being published in 1961 as the lead article in the November issue of the journal *Arbor*,[13] Ibáñez Martín, now the Spanish ambassador to Portugal, once again and openly reiterated that the reason for the creation of the CSIC had been none other than 'to inject vital enthusiasm into the work of those who were not satisfied with the moral and material poverty of Spain in 1936'.[14]

Almost a quarter century after the end of the Spanish Civil War, Manuel Lora-Tamayo, secretary-general of the Juan de la Cierva Board for Technical Research and now Minister of National Education, also insisted on the existence of 'a new scientific flourishing' in Franco's Spain: 'In addition to publications in our own language, many Spanish scientists now publish in foreign journals . . . Today . . . in the annual indices of a great variety of journals . . . Spanish surnames can be happily found' (Lora-Tamayo, 1963: 21). In this narrative, names served as a tangible example that the new Spanish science was beginning to be valued abroad, although in practice, he said, 'only the same dozen names of the last 30 years continue to be bandied about'.[15]

Some of the participants in this 'scientific flourishing' – who, according to Lora-Tamayo, were allegedly overlooked – have appeared in this book: not only male scientists from the Juan de la Cierva Board like Juan M.ª Torroja, José Luis Amorós, José María Garrido, Eduardo and Agustín Vioque Pizarro and Manuel Lora-Tamayo himself, but also female scientists such as Clara Díez de Bethencourt, Olga García Riquelme, Laura Iglesias and Juana Bellanato. These were only some of the authors – mostly, but not only men – who in the late 1940s and, primarily, 1950s published in other languages and in foreign scientific journals at a time when English was beginning to dominate the international scientific vocabulary in the face of the decline of German and French, riding on the back of the scientific and political predominance of the United States (Gordin, 2015: 293–316). Although these names were presented as achievements of the Franco regime, they were largely individual cases associated with specific figures who had institutional sway. However, this never prevented them from being used as arguably irrefutable proof of the perpetual connection between Franco's Spain and the international scientific community (García Naharro, 2020b).

Part IV

Places

Chapter 13

The CSIC Libraries: Library Services and International Exchange

'Science will be properly established in a setting that reflects the interest and importance that the Government attaches to its work.'[1] With these words, spoken at the opening of the sixth plenary session of the Spanish National Research Council in 1945, José Ibáñez Martín was referring to the intention expressed by the dictatorship to provide Spanish science with facilities commensurate with its institutional significance. Since 1942, he added, this scientific endeavour had been

> 'renovating its facilities, constructing new ones and providing the Fatherland with scientific material, resources and technical and intellectual results that once had to be sought outside the country, in a studied display – malicious in some of its aims – of ignorance of, or contempt for, our own science'.[2]

Once more, the official discourse of post-Civil War Spain did not miss an opportunity to brand the recent past as pro-foreign, while the Franco regime took on the arduous task of national restoration, cemented both rhetorically and materially in a combination of reconstructions and constructions from scratch. All of this, of course, was an integral part of the programme to highlight the regime's civil works which, as with the inaugurations of the CSIC facilities, always received extensive media coverage (Ramírez Martínez, 2011: 163–82). Like the central CSIC building, these constructions materially incorporated an imperial rhetoric, based on the mere provenance of their foundations, thus, in the words of Ibáñez Martín, extracting 'noble symbolism from the rich wealth of the stony ground of Spain, a reflection of the rugged steadfastness of the ideals of the Council, the marble-like, cold objectivity of its technologies and the polychrome complexity of its enterprises'.[3]

The work entrusted at the time to the CSIC was in no way easy. The council was the scientific institution *par excellence* of Franco's Spain, the 'work of men of faith in Spanish science' and the 'core of the Spanish resurgence',[4] 'born out of the essential work of research fellows and culminating in the rigorous preparation of professional researchers'.[5] To conduct these many important activities, in turn, the professional researcher required specific places, not only 'the classroom, the seminar and the laboratory',[6] but also, as Ibáñez Martín stressed, 'the quiet enclosure of the library',[7] where the researcher's knowledge came face to face with the repertoire of learning, in other words, the indispensable condition of science: the international communication and exchange of knowledge. The CSIC was always aware of this inevitable requirement as was, by extension, the Franco regime. Beyond the ideological dictates of its official discourse, the dictatorship worked from the very beginning – as determined by the available means and its own interests – to provide the council and its centres with 'the books and journals that bring information about what is being done beyond our borders to our men'.[8]

Accordingly, the task of reconstructing scientific knowledge in Spain under Franco must also account for the work done by the council's bibliographic services, then structured around the CSIC General Library. One of the 'Libraries of high culture in Spain', in the words of Amadeo Tortajada Ferrandis, the head CSIC librarian since 1942, its collection – combined with the specialist libraries in the council's various institutes – held nearly a million works by the late 1940s.[9] However, as recently as October 1940, this library, the so-called council library, did not even exist, and neither did the main council building itself, although plans to install the library on the ground floor of the new edifice were in place.[10]

It was actually not until 1944 that both the collection and the 'central CSIC services, including the main library with all the bibliographic services' were finally given a home in the new, as yet unfinished, central CSIC building located near the former National Institute of Physics and Chemistry building, which now housed the Alonso Barba and Alonso de Santa Cruz Institutes.[11] Therefore, lacking buildings of its own, the CSIC General Library ended up occupying the right wing of the council's main headquarters, in an emblematic building at 117 Serrano Street in Madrid that was solemnly inaugurated on 12 October 1946. As Amadeo Tortajada recalled in the late 1940s:

> The library currently occupies three floors, a repository and several working departments. On the main floor, on the right, are the librarian's office, the Cataloguing and Classification sections, the Bibliographic Information and Catalogues, the reference work and consultation reading room and the foreign journal room; on this floor on the left are the offices of the International Department of Modern Cultures and the Journal Cataloguing section . . . The Department of International Exchange has been set up in the right wing of the building's main floor.[12]

The collection was largely supplied by exchanges with other cultural institutions. While few and far between in the early years due to 'the disturbing, external state of abnormality established by the [Second World] War', beginning in 1945 the exchanges increased,[13] placing, according to the CSIC, 'in our hands each year, 5,000 journals and several thousand works published in the United States, France, England, Portugal, Canada, Latin America, Sweden and other countries'.[14] To assist in this process, the CSIC General Library also 'benefitted from duty-free treatment, exempt from the previous import-export license',[15] which was required of other private publishers and booksellers in the Spanish publishing field at the time. The Franco regime did not only use these legal and regulatory privileges to acquire national and international publications as a barometer of the growing importance of the institutional scientific sector to the dictatorship, but also, it was assumed, to reflect the unbreakable bond between the Spanish scientific community and the outside world. Writing on this subject in May 1943 to the director of the CSIC Institute for Physics and Chemistry, Amadeo Tortajada reported that:

> This General Library has the ability to quite quickly obtain French and American books, according to the agreement made. To that end, we have appointed a representative in these countries and since a general order is being prepared to deliver shortly, we ask you to draw up for this [International Exchange] office a list of what French and American books you desire so as to order them immediately.[16]

Books, journals and other foreign scientific publications came from a variety of sources: the Government of France, where Professor Aurelio Viñas was the delegate appointed to acquire the books,[17] the Government of the German Third Reich and the Rockefeller Foundation in America, the council libraries through Spanish booksellers or distributors with permission to import, as well as the collections of the former Board for Advanced Studies and Scientific Research (JAE), confiscated libraries and the sale of individual library collections (García Naharro, 2015b: 769–71). Additionally, the CSIC's own Publications Office gave each of the council's institutes two issues of the journals and books published under the CSIC's auspices for their libraries, requesting, in turn, that they send the Publications Office 'the total count of the printed copies corresponding to the works published by them'.[18]

Moreover, the Order of 18 February 1941 stipulated that all middle and secondary education schools and 'University, Provincial and Public Libraries' in Spain acquire CSIC publications on a compulsory and 'preferential basis above all others'. These acquisitions were paid for by 'the subsidies that these Schools and Establishments receive from the General State Budget or funds from their own fundraising for scientific and educational material'. The Order stated that the aim was none other than 'to keep the Teaching Staff, constituents of the school and public in contact with the work being done by the various Institutes of the Spanish National Research Council and its bibliographic orientation'.[19]

All things considered, the preferred physical locations to consult the publications of the CSIC were its various libraries, from the General Library to the specialist libraries in the institutes, where they shared space with scientific works and periodicals from abroad. These were, in short, scientific libraries, places of learning, silent and sheltered, isolated and a far cry in outward appearance from what was occurring outside their walls. However, the CSIC libraries were never cut off from the time passing around them, but stayed involved through, one might say, a thousand intermediaries (records, files, slips, reports, evaluations, photographs, inventories) gathering information about 'the battle that rages outside', whose overall meaning would be lost without this panoptic view and notarial compilation (Latour and Hermant, 1999: 181). Of all these scientific libraries, one of the most important, without question, was the collection housed in the striking former Rockefeller Building, which came to be officially known in 1946 as the Antonio de Gregorio Rocasolano Institute for Physical Chemistry.

Chapter 14

The Alonso Barba and Alonso de Santa Cruz Institutes Library: The Antonio de Gregorio Rocasolano Institute for Physical Chemistry

It was from the embryonic Blas Cabrera y Felipe Physical Research Laboratory that the JAE's National Institute of Physics and Chemistry emerged in 1932, located in what was colloquially known as the Rockefeller Building. A place and institution that was rife with history, however, lost its name with the creation of the CSIC under Franco (although it continued to be used in much of the correspondence of the time), and came to house the Institute of Physics / Alonso de Santa Cruz Institute and the Institute of Chemistry / Alonso Barba Institute, both directed by José Casares Gil, with Julio Palacios and Antonio Rius, respectively, as associate directors.

There, in a building originally paid for by the Rockefeller Foundation, the library for both institutes was then installed (although it had been in operation since the time of the Blas Cabrera y Felipe Laboratory in the Palace of Industry, located in the Altos del Hipódromo suburbs of Madrid), and Casares Gil began the complicated task of renewing subscriptions to foreign journals and replacing the lost issues (Iglesias Sánchez, 2009: 219–22). This was truly necessary because, as Palacios wrote to Paul Scherrer in June 1939, at that time the library did not have 'a single book or journal of those published during the War[1] and . . . It will be very difficult for us to resume our work if we do not receive generous assistance from friendly nations, and get them to send professors, books, journals and scientific material to us' (Fernández Terán, 2014: 562). In May 1940, the director of the Institute of Physics and Chemistry repeated that sentiment in a letter sent to the undersecretary of the Ministry of National Education:

> The Library at this Institute, which owes all its importance to its excellent collection of physics and chemistry journals, suffered a serious loss during the Marxist domination, because they stopped receiving almost all the foreign publications produced during that time. Filling the gaps created in its collection for that reason would involve a considerable amount of foreign money and that is why I wrote the Rockefeller Foundation to ask them to address our need. The response was fully satisfactory, as they offered to supply us with the amount we need in dollars for this purpose.[2]

Although the geopolitical conditions of the Second World War were not the most propitious, as an official institution of the Franco regime, the library of the Alonso Barba and Alonso de Santa Cruz Institutes at that time had various sources of assistance. The German, French and Italian governments[3] all agreed to complete the collections lost during the Spanish Civil War, with the German ambassador offering in January 1940 to supply the collections, along with those of the CSIC general library and other council institutes, with the German journals published between 1936 and 1939.[4] Thus, around April 1941, the German Book Exchange Centre in Berlin sent through Rudolf Kadner, a bookseller with close ties to the German embassy in Madrid, a significant collection of German scientific journals to various Spanish research centres.[5]

However, during this time, some journals, such as *Annalen der Physik, Helvetica Physica Acta, Proceedings of the Physico-Mathematical Physical Society of Japan* and *Science Reports of the Tohoku Imperial University* apparently continued to arrive, as noted by the Rockefeller Foundation in the United States, which also expressed its willingness to complete the list of twenty journals initially requested by Professor Miguel Catalán – the number increased to fifty-two after Julio Palacios made his request – as long as they received the consent of the minister of national education.[6] An offer from Argus Suisse et International de la Presse S.A. in Geneva to send 'newspaper and journal articles from all the requested countries, noting the date and title of the paper that the requested article was cut out of' was even rejected in December 1939. The director of the (still called) National Institute of Physics and Chemistry answered that 'We do not deem the offer from this Centre convenient at this time, bearing in mind that, thanks to the support of the Undersecretary [of the Ministry of National Education], we hope shortly to complete our collection of foreign journals, with which we are satisfied and very thankful'.[7]

Paradoxically, during the early 1940s – the years of the 'reactionary utopia of the victorious National Catholicism in the Civil War' (Canales, 2019: 40–6 and 2015: 79–102), while the regime's official discourse was urging Spanish scientists to forget the era of submission to foreign postulates and revive the splendour of Spanish scientific tradition, the scientists, on the other hand, were demonstrating a pressing need to maintain contact with what was happening abroad. Moreover, after 1939, the Ministry of National Education strove – within its limitations – to acquire all the books and journals that the laboratories required, whether because 'they had disappeared due to the War or because they were necessary to run the service'.[8] Additionally, foreign institutions like the American Chemical Society, the publisher of the *Journal of the American Chemical Society*, *Chemical Abstracts*, *Industrial and Engineering Chemistry* and *News Edition*, continued to send information and subscription prices for these journals from abroad to the CSIC's National Institute of Physics and Chemistry during the 1940s.[9]

However, despite these efforts, journals like *American Physics Teacher* had not been received by the National Institute of Physics and Chemistry library since June 1936, and the attempts made to complete the collection by the director of the institute with the CSIC and by Julio Palacios with the secretary of the Book Acquisition and Exchange Board, Federico Navarro Franco, were still unsuccessful as of May 1941.[10] The same applied to other important journals like *Annals of Science, The Analyst, Annales de Physique* and *Annales de Chimie*. Moreover, in June 1940, the National Bureau of Standards informed the library that, due to the circumstances of the Second World War, the international exchange service of the Smithsonian Institution had been interrupted, and the library would have to wait to complete its collection of *Journal of Research*.[11] Such was the state of things when the Exchange Section asked the various CSIC institutes for a general list of journals that they were receiving. The results provided in May 1941 by the library of the National Institute of Physics and Chemistry – now the Alonso Barba and Alonso de Santa Cruz Institutes Library – were not encouraging.[12]

Table IV.1. List of journals received by the Alonso Barba and Alonso de Santa Cruz Institutes Library before the 'Movement' that continue to arrive at this time (May 1941)

Journal	Publisher	Exchange or subscription
Anales de la Real Sociedad de Física y Química	RSEFQ	
Annales de Chimie	Several	Exchange
Annales de Physique	Several	Exchange
Annalen der Chemie	Wisland	Subscription
Annalen der Physik	Der deutschen chemischen Gesellschaft	Subscription
ATM	F. Müller	
Berichte der deutschen chemischen Gesellschaft	Der deutschen chemischen Gesellschaft	Subscription
Chemiker Zeitung	Otto von Halem	Subscription
Chemisches Zentralblatt	Der deutschen chemischen Gesellschaft	Subscription

Helvetica Physica Acta	Schweizerische physikalische Gesellschaft	
Investigación y Progreso		
Journal of Science of the Hiroshima University	Hiroshima University	Exchange
Kolloid Chemische Beihefte	W. Ostwald	
Die Naturwissenschaften	Several	Subscription
Physikalische Zeitschrift	Der Physikalischen-Technischen Reichsanstalt	
Proceedings of the Physico- Mathematical Society of Japan	The Physico-Mathematical Society of Japan	
Science Reports of the Tohoku Imperial University	Tohoku Imperial University	
Zeitschrift für Analytische Chemie	H. Fresenius	
Zeitschrift für Instrumentenkunde	Der Physikalischen-Technischen Reichsanstalt	
Zeitschrift für Anorganische Chemie	Several	
Zeitschrift für Elektrochemie	Der deutschen Bunsen Gesellschaft	
Zeitschrift für Physik	Der deutschen Physikalischen Gesellschaft	Subscription
Zeitschrift für Physikalische Chemie	Several	
Zentralblatt für Mathematik	Der deutscher Mathematiker Vereinigung	Subscription

Table IV.2. List of journals received by the Alonso Barba and Alonso de Santa Cruz Institutes Library that began to arrive with the new operation of the institute

Journal	Publisher	Exchange or subscription
Anales del Jardín Botánico de Madrid	CSIC	Exchange
Anales de Ciencias Naturales	CSIC	Exchange
Afinidad	Asociación de Química de Sarriá	Exchange or gift
Combustibles	CSIC	Exchange
Educación	Ministerio de Educación Nacional	
Eos. Revista Española de Entomología	CSIC	Exchange
Psicotecnia	Instituto Nacional de Psicotecnia	Subscription
Rediconti del Seminario della Facultá di Scienze della R. Universitá di Cagliari	Facultá di Scienze della R. Universitá di Cagliari	
Revista Askania	Askania Werke	Gift
Revista de la Universidad de Tucumán	Universidad de Tucumán	Exchange
Revista Ibérica de Parasitología	CSIC	
Revista de la Facultad de Química industrial y agrícola. Anales del Instituto de investi-gaciones científicas y tecnológicas	Instituto de Investigaciones Científicas y Tecnológicas	Exchange
Trabajos del Laboratorio de bioquímica y química aplicada	CSIC	Exchange or gift

Table IV.3. List of journals once received by the Institute of Physics and Chemistry (Alonso Barba and Alonso de Santa Cruz Institutes) that are no longer received

Journal	**Publisher**	**Exchange or subscription**
Anales de la Asociación Española para el Progreso de las Ciencias	Asociación Española para el Progreso de las Ciencias	
Anales de la Asociación de Antiguos Alumnos del ICAI	Asociación de Antiguos Alumnos del ICAI	Exchange
The Analyst	The Society of Public Analysts	
Angewandete Chemie	Verein deutscher Chemiker	Subscription
Annais da Facultade de Ciencias do Porto	Facultade de Ciencias do Porto	Exchange
Annals of Science	D. Mckie, London	
The American Physics Teacher	The American Institute of Physics	
The Astrophysical Journal	H. C. Cale, Chicago	
Annales de Chimie Analytique	Several	
Boletín de la Academia de Ciencias	Academia de Ciencias	
Bureau International de Poids et Mesures	Comité International des Poids et Mesures	Exchange
British Chemical Abstracts	The Chemical Society, London	
Bulletin de la Societe Chimique de France	La Societe Chimique de France	

Collection de Travaux Chimiques de Tchecoslovaquie	Societas Scientiarum Bohemica	
Communications from the Physical Laboratory of the University of Leiden	Universiteit Leiden	Exchange
Comptes Rendus des Seances de l'Academie de France	L'Academie de Sciences	
Chemical Abstracts	The American Chemical Society	Subscription
The Chemische Fabrik	Verein deutscher Chemiker	
Chemie et Industrie/ Chimie Industrielle	Societé de Chimie Industrielle	
Giornale de Chimica Industriale ed Applicata	Sociedad de Química Industrial, Milán	
Olas und Apparat	R. Wagner Sohn	
Helvetica Chimica Acta	Schweizerische Chemische Gesellschaft	
Indian Journal of Physics	The Indian Association for the Cultivation of Science	Exchange
Industrial and Engenieering Chemistry	The American Chemical Society	Subscription
Journal of the Acoustical Society of America	The American Institute of Physics	
Journal of the American Chemical Society	The American Chemical Society	Subscription
Journal of the Chemical Education	The American Chemical Society	

Journal of the Chemical Physics	The American Institute of Physics	
Journal of the Chemical Society	The Chemical Society, London	
Journal of the Optical Society of America	The American Institute of Physics	
Journal of the Physical Chemistry	The American Institute of Physics	
Journal of Scientific Instruments	The Institute of Physics, London	
Journal de Chimie Physique	The Societé de Chimie Physique	
Journal de Physique et le Radium	Societé Francaise de Physique	Exchange
Journal of Research	The National Bureau of Standards, Washington	Exchange
Nature	MacMillan and Co., London	
Il Nuovo Cimento	Societa Italina de Física	
Philosophical Magazine and Journal of Science	Sir Oliver Lodge	
Proceedings of the Physical Society of London	The Physical Society of London	
Proceedings of the Royal Society	The Royal Society, London	
Proceedings of the Koninklijke Akademie te Amsterdam	The Royal Academie of Amsterdam	Exchange
Publications du Laboratoire de Physique de l'Université de Coimbra	Universidade de Coimbra, Portugal	Exchange

The Physical Review	The American Institute of Physics	
Reveu d'Acoustique	Instituto Henry-Poincaré	
Revista de Ingeniería Industrial	Asociación Central de Ingenieros Industriales	
Recueil des Travaux Chimiques de Pays-Bas	Nederlandsche Chemische Vereeniging, Amsterdam	
Revista de la Academia de Ciencias Exactas, Física y Naturales	Academia de Ciencias Exactas, Físicas y Naturales	
Review of Modern Physics	The American Institute of Physics	
Review of Scientific Instruments	The American Institute of Physics	
Science Abstracts	The Institute of Electrical Engineers, London	
Transactions of the Faraday Society	The Faraday Society, London / Edinburgh	
Physikalische Zeitschrift der Sowjetunion	Volkskommissariat für Schwerindustrie der USSR	Exchange
Zeitschrift für Kristallographie	Several	

'The publisher informed us today that they are completely out of those issues. We are very sorry that we cannot oblige you.'[13] Thus did the German bookseller Rudolf Kadner based in Madrid (17 Serrano Street) communicate to the library that, in June 1943, issues 2–4, 7–10, 12–15, 17–23 (1941) and issue 6 (1940) of the journal *Die Naturwissenschaften* were still unavailable, as would be the case with the 1942 issues, according to the Herder bookshop, which would never again be available as a complete set.[14] That same year, Rudolf Kadner also worked with the library to renew its subscription to *Die Naturwissenschaften*, in addition to other German publications like *Berichte der deutschen chemischen Gesellschaft*, *Zentralblatt für Mathematik*, *Chemiker Zeitung*, *Annalen der Physik*, *Kolloid-Beihefte Zeitschrift* and *Chemisches Zentralblatt*.[15]

Against all odds, by 1942 the situation in the library had significantly improved; the Book Acquisition and Exchange Board had completed its collections of journals like *Proceedings of the Royal Society of London* and, thanks to the French Institute and the bookseller Rudolf Kadner, the status with other publications like *Journal de Chimie Physique*, *Journal de Physique et le Radium*, *Chemisches Zentralblatt*, *Zeitschrift für Anorganische und allgemeine Chemie* and *Die Chemische Fabrik* had been restored.[16] On 27 October 1943, Amadeo Tortajada Ferrandis wrote to the secretary of the Alonso Barba Institute to tell him that he could stop by the International Exchange Section at 121 Serrano Street 'any afternoon from 4 to 8' to pick up 'the journals *Journal of the American Chemical Society* (vols. 60 to 64), *British Chemical Abstracts* (1935–42) and *American Chemical Abstracts* (1938, 1939 and 1940)'.[17] One year later, they had also made advances towards completing the collection of the journals *Nature* (1938–42), *Journal of Scientific Instruments* (1926–7, 1936–1940), *Annals of Science* (1936–41), *The Analyst* (1936, 1941 and 1942), *Physics Abstracts* (1941 and 1942), *Science Abstracts* (1937–42) and *Transactions of the Faraday Society* (1937–42),[18] which, combined with the other journals in the library, meant that they had a total of 3,787 volumes in 1944. According to an internal library report from March 1945, they were 'currently subscribed to 12 German journals through the Kadner bookseller. We are receiving the other German, French, English and American journals through the Spanish National Research Council Exchange Section'.[19]

Along with the incessant requests for journals made to the International Exchange Section in the early 1940s – issues of *The Analyst*, *Annals of Science*, *Journal of Scientific Instruments*, *Nature*, *Physics Abstracts*, *Science Abstracts* and *Transactions of the Faraday Society*,[20] among many others – the Alonso Barba and Alonso de

Santa Cruz Institutes library also requested books of interest in English, German and French. These ranged from the monographs *Einführung in die Quantentheorie der Wellenfelder* by the German physicist Gregor Wentzel (1898–1978) to *Kernphysikalische Tabellen* by the nuclear physicist Josef Mattauch (1895–1976)[21] and *Atoms in Action: The World of Creative Physics* and *Thermodynamics and Chemistry* by the American physicists George Russell Harrison (1898–1979) and F. H. MacDougall (1883–1974).[22] These titles, requested at the height of the Second World War, clearly demonstrate an interest in the international field of physics and chemistry at that time, subjects that had also been of great interest to two of the most well-known Spanish scientists in their day who had, however, been consigned to oblivion under Franco: Blas Cabrera and Enrique Moles.

Chapter 15

Confiscated Collections: The Private Libraries of Blas Cabrera and Enrique Moles

> In June 1940, the Library had 871 volumes of books arranged on the shelves by topic . . . There was a card index by author, subject and book title . . . without any further information . . . There were also loan slips, only by book title . . . The former Director's Office had a Physics Library made up of 749 volumes that I found completely disorganized, with a card index that was useless for the same reasons as that of the large Library.[1]
>
> María Fernández Vallespín, librarian, Alonso Barba and Alonso de Santa Cruz Institutes Library

The books that the librarian describes in the Physics library may well have come from the private library of Blas Cabrera, the former director of the JAE's National Institute of Physics and Chemistry, who kept some of his personal collection at the institute. The remaining texts were also transferred there under Franco, 'apparently to be stored and safeguarded' according to Blas Cabrera's son, the architect Luis Cabrera Sánchez, who declared on 30 November 1945 that the Alonso de Santa Cruz Institute Library made him hand over 'some of the books that comprised the Library of my deceased father, Blas Cabrera (RIP)'.[2] From Mexico, exiled Spanish chemist Antonio Madinaveitia pondered his loss:

> On the first of August 1945, Blas Cabrera died in Mexico. Spain is losing in exile the men who promoted the intense scientific development that occurred there from the beginning of the century to

> the [Franco's] military uprising. José Castillejo has died; Ignacio Bolívar has died; and now Cabrera.[3]

Born in the Canary Islands and a regular at Santiago Ramón y Cajal's social gatherings at the Café Suizo in Madrid, Blas Cabrera y Felipe (1878–1945) became a chair professor at the Universidad de Madrid in 1905, director of the Laboratory for Physics Research at the Board for Advanced Studies and Scientific Research (JAE) in 1910 and member of the International Committee of Weights and Measures in 1931. Unfortunately, although these credentials made him an eminent physicist in Spain and brought him international recognition for his contributions to the field of magnetochemistry, this did not prevent him from forming part of the Republican exile after the outbreak of the Spanish Civil War (Romero de Pablos, 2002: 15–64; Sánchez-Ron and Roca-Rosell, 1993). Thus, while his circumstances in October 1936 took him first to France and then to Mexico, his library remained in Madrid, a witness to and faithful reflection of both his research interests and his conviction about the crucial importance of quantum theory and research into the structure of the atom for physics and chemistry at that time.

According to the lists signed by Luis Cabrera Sánchez, this material legacy included some 835 titles[4] on topics related to physics and chemistry published in a variety of languages, with a clear predominance of French as the original or translated language (44.8 per cent). This compendium of highly valuable works by internationally renowned authors included eight texts in French (one volume without a cover) by the acclaimed French mathematician and physicist Henri Poincaré (1854–1912),[5] ten titles in various languages by the German theoretical physicist and Nobel Prize recipient Max Planck (1858–1947),[6] eight titles by two other Nobel Prize recipients, Joseph John Thomson (1856–1940)[7] and Louis de Broglie (1892–1987),[8] and nine by Sir Arthur Eddington (1882–1944).[9] The collection also contained seven titles in French by the physicist Henri Bouasse (1866–1953),[10] one of the most often-cited anti-relativists, which shared space on the lists with works by Albert Einstein (1879–55),[11] James Hopwood Jeans

(1877–1946)[12] and Max Born (1882–1970),[13] some of the fathers of the New Physics.

Figure 1. Number of titles by language in Blas Cabrera's private library

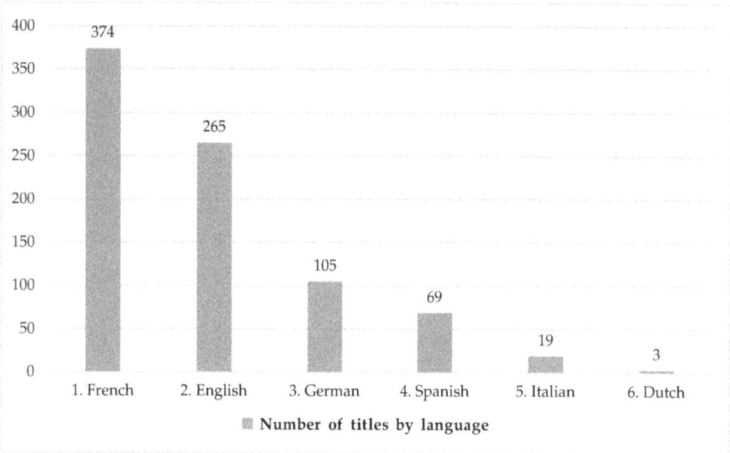

Source: Author

Many of these works were returned to Blas Cabrera's heirs in late November 1945, officially 628 books, although the actual number was slightly lower and a few more were missing, as Luis Cabrera Sánchez later confirmed when he urged the CSIC to rectify the absence and 'investigate who is in possession of these books and demand that they be returned'.[14] The missing books included the doctoral thesis by Dutch physicist Hendrika Johanna van Leeuwen (1887–1974) and books by the British mathematician Ebenezer Cunningham (1881–1977), the Indian astrophysicist Meghnad Saha (1893–1956), the German physicist Arnold Sommerfeld (1868–1951) and even publications by the Solvay International Institute for Physics on whose scientific board Blas Cabrera served, on the recommendation of Marie Curie and Albert Einstein (Romero de Pablos, 2002: 57–59). Of the books requested by Cabrera's heirs, several were indeed returned, either formally or informally.[15] Others were never recognised as belonging to him, but were rather considered the property of the institute. Most of the books requested from the CSIC were published in English (48.4 per cent), although they also included some written in Spanish by colleagues at the Free Institution of Education (ILE), like Domingo de Orueta y Duarte (1862–1926), and former colleagues at the JAE's National Institute of Physics and Chemistry who were now members of the CSIC, such as Julio Palacios (1891–1970).

Of the erstwhile scientists at the Board for Advanced Studies and Scientific Research, another salient scientific figure was the head of the chemistry section at the Rockefeller Building and chair professor of inorganic chemistry at the Universidad de Madrid, Enrique Moles Ormella (1883–1953). After receiving his doctorate in sciences from the Universität Leipzig and being instructed in research by Philippe A. Guye (1862–1922) at the Université de Genève, Moles achieved international recognition for the determination of atomic weights. His achievements in research were matched by his accomplishments in education, where he modernised teaching in his discipline. However, after he took the side of the legitimate Republican government during the Spanish Civil War, he was expelled from his professorship and the National Institute of Physics and Chemistry and went into exile in Paris. On his rash return to Spain after the war, some of his former colleagues denounced him and he was subsequently sentenced to a year and a half in Porlier Prison in Madrid. When he was released in December 1943, he was stripped of all his positions and condemned to institutional marginalisation until the end of his days (Giral, 1994: 108–17; Romero de Pablos, 2002: 65–110; Sales, 2021; Sales and Nieto-Galan 2014 and 2019). By then, moreover, one of his most valuable possessions had been taken away from him: his library.

In a letter dated 20 June 1939, the director of the National Institute of Physics and Chemistry wrote:

> Having heard that the library of the former professor of this Institute, Enrique Moles, located at 21 Jordan Street, 1st floor, right, had been sealed off by the Requisition Council of the Army First Corps, and considering that the books, because of their special

> content, would be of great service to the library in this Institute, where they would be duly safeguarded, I ask you, if you consider it desirable, to arrange or authorize their transfer, upon inventory or any necessary formalities.[16]

The request on behalf of the institute received a response a few days later when, after the military command consulted the Requisition Council, the director was informed that the decision would be made by the special judge for political responsibilities,[17] who ultimately approved the request to transfer the books.[18] This was confirmed in February 1940 by the head of the Library Recovery Service, Vicente Navarro-Reverter, who noted that 'amongst the books of Mr Moles that were deposited with this Institute' were nine that belonged to the Faculty of Pharmacy library, 'easily identifiable because they have an ownership stamp',[19] that needed to be returned.[20] These were actually historical texts, three written by the French chemist Louis Proust (1754–1826)[21] and six by his disciple, the Catalan Francisco Carbonell y Bravo (1768–1837),[22] once lent by the pharmacy library to Professor Moles. In addition to the requested works, the director of the National Institute of Physics and Chemistry discovered among Enrique Moles's books several more with the stamp of the Faculty of Pharmacy and offered to collaborate to complete the collections with missing works.[23]

The director of the National Institute of Physics and Chemistry was not mistaken in believing that, thanks to their contents, the books in Enrique Moles's private library would be extremely valuable for the Institute. Moles's collection contained some 564 scientific titles in more than 600 copies,[24] in addition to twenty-four filing cabinets containing offprints on physical chemistry and inorganic chemistry, twelve filing cabinets with works by Moles and his collaborators and several collections of scientific journals, including *Zeitschrift für Physikalische Chemie* (volumes 1 to 88), *Journal de Chimie Physique* (volumes 14 to 21 and 33), *Zeitschrift für Elektrochemie* (volumes 1 to 22), *Journal of the Physical Chemistry* (volumes 1 to 18) and *Anales de la Sociedad de Física y Química* (volumes 11 to 33).[25] These journals, documents and books may have ended up being stored at the time in the room annexed to the director's library, which was found in 1945 to contain 'a collection of physics booklets, all predating 1937, and some extremely incomplete journals'.[26] Be that as it may, both these and other journals that belonged to Moles – for example, the *Revista de la Residencia de Estudiantes* and *Boletín Tecnológico* – were finally returned to him in November 1948, along with nearly 400 additional printed products, thus righting the wrong committed almost a decade earlier, albeit only partially.[27]

Like the handful of books claimed by the Faculty of Pharmacy that Moles kept to consult when necessary, the works from his private library make it possible to reconstruct part of his intellectual evolution, as well as the sociabilities and exchanges of his past (Chartier, 2018: 28–40). The books from the pharmacy library alone chronicle his education and his successful pharmacy studies in Barcelona, which culminated in his doctoral degree from the Universidad Central in 1906. Other works speak to the Universität Leipzig and the institute of German chemist and Nobel Prize winner Wilhelm Ostwald (1853–1932), the father of physical chemistry, along with Svante Arrhenius (1859–1927) and Jacobus Henricus van't Hoff (1852–1911); the existence of sixteen books in the inventory by Wilhelm Ostwald and several editions of *Hand und Hilfsbuch der physiko-chemischer Messungen* by Ostwald and Robert Thomas Diedrich Luther clearly attest to this.

Writing in 1910, Moles observed that 'all, or almost all, the current Physical Chemistry leading figures passed through the Ostwald Institute' (Moles, 1911: 70), including several under whom he studied in Leipzig, and whose books he also kept: Max Le Blanc (1865–1943), Carl Drucker (1876–1959), Herbert Freundlich (1880–1941) and Wilhelm Carl Böttger (1871–1949), the last of whom, according to Moles, enjoyed 'the reputation of an exquisite and extremely exact analyst, in Leipzig, across Germany and even abroad'.[28] Moles's time in Germany was also represented by reference texts like *Vorträge über physikalische Chemie* (1907 edition) by Ernst Cohen (1869–1944), *Einführung in die mathematische Behandlung der Naturwissenschaften* (1904) by Walther Nernst (1864–1941) and Arthur Moritz Schoenflies (1853–1928) and multiple copies of books by influential figures like the

Dutch van 't Hoff (1852–1911), the first winner of the Nobel Prize in chemistry, as well as the biography of van 't Hoff written by his disciple Ernst Cohen.

Ernst Cohen was one of the professors invited by Moles to act as a special advisor on the study committee at the first International Chemistry Meeting at La Magdalena (Santander), held in 1933. This meeting was attended by two Nobel Prize winners, Fritz Haber (1868–1934) and Richard Willstätter (1872–1942), whose books Moles also had in his library (one of them, *Hormonas, vitaminas, fermentos* (1934), was the product of the inaugural International Santander Summer School of 1933). One year later, Moles met up with the German chemist Hans Georg Grimm (1887–1958) in that same programme (Madariaga de la Campa and Valbuena Morán, 1999), and later added to his private library a copy of the book written by the German, *Química y estructura atómica* (1936), a product of that course. These books, in and of themselves, serve as a material reflection of the efforts made during the Second Republic to internationalise Spanish science and establish contacts with the most esteemed representatives of science from around the world (Nieto-Galan, 2019: 65–73) (López-Ocón Cabrera, 2023).

Figure 2. Number of titles by language in Enrique Moles's private library

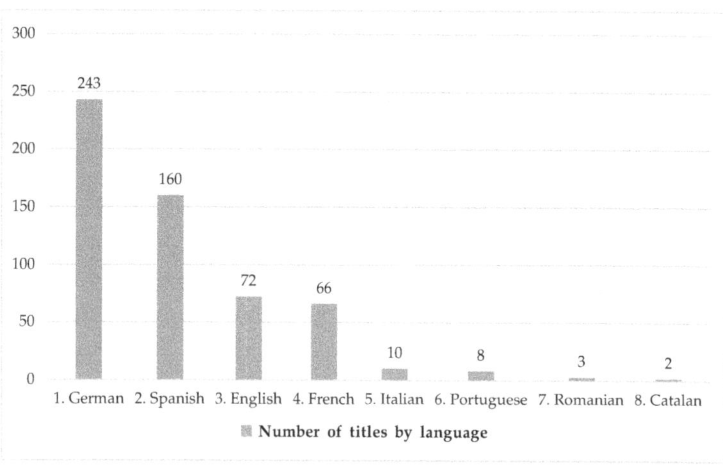

Source: Author; multiple editions of certain works are represented

Moles's international experience was also reflected in his private library. Unlike Blas Cabrera, for instance, his library was dominated by original works published in German (43.1 per cent) and English (28.4 per cent), languages that Moles considered indispensable for the study of chemistry: 'At the present time, to learn Chemistry, if you can speak English, you are not completely defenceless. However, for an advanced study that is both rigorous and serious, German is an indispensable tool.'[29]

Figure 3. Number of titles by publication date in Enrique Moles's private library

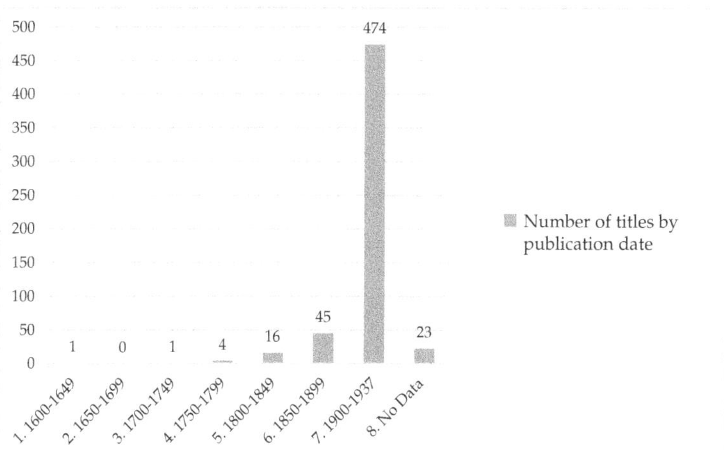

Source: Author; multiple editions of certain works are represented

Taken together, the private libraries of Blas Cabrera and Enrique Moles constitute the thematically specific bibliographical inventories that profile the identity of each scientific discipline (Nye, 1993: 13–31), incorporating illustrious names, benchmark institutions and subjects of interest that function here as a kind of genealogy inscribed in works written in diverse languages by authors of different nationalities. This compendium of printed products – journals, booklets, reports, annual reports, books and manuscripts – certified by scientific societies, universities and research centres, accompanies the names of the eminent men of international physics and chemistry, among whom the name of a woman, Marie Curie (1867–1934), shone in its own right in Moles's library: Mme. Curie, 'so that no one confuses her with M[essieur]. Curie' (Roqué, 2011: 16) and her book *L'Isotopie et les éléments isotopes* (1924).

When Curie's text was published, Albert Einstein was already a mythical figure and the symbol of science itself (Glick, 2005: 15), as was his book *Über die spezielle und allgemeine Relativitäts-theorie* (1916), of which Enrique Moles had an edition from 1921. The other key reference works from the new physics in his library included *Acht Vorlesungen über theoretische Physik* (1909) by Max Planck, Xavier Zubiri's translation of the work by Schrödinger, *La nueva mecánica ondulatoria* (1935) and the first edition of *Principio de la relatividad* (1923) by his colleague Blas Cabrera, the leading figure on the periphery of European science, published in Madrid by Residencia de Estudiantes.[30]

The most notable works in the Spanish language in Moles's library were written by his Spanish teachers: the biochemist, rector and member of the Atheneum of Madrid, José Rodríguez Carracido (1856–1928) and the chemist and pharmacist José Casares Gil (1866–1961), director of the CSIC Institute for Physics and Chemistry, a friend in years gone by and early protector – whose name, however, appears among the illustrious colleagues who spoke out against Moles during the long trial that culminated in his conviction and disqualification in 1942 (Sales and Nieto-Galan, 2014: 159–61; Sales, 2021: 293–4).

Finally, Moles's private library contained seven copies of the works of Modesto Bargalló (1894–1981) – the 'eminently learned Professor at the Escuela Normal de Guadalajara', as Moles described him in a review of the second edition of *Manual de Química* (1927), which 'left him highly impressed' – who had been living in exile in Mexico since February 1939 (Moreno Martínez, 2019: 246); *Física y Química* (1934) by Professor Miguel Catalán and Professor Andrés León, both removed from their positions after the Spanish Civil War; and *Tratado de Bioquímica* (1928) by the chair professor of chemistry at the Universidad de Zaragoza, Antonio de Gregorio Rocasolano (1873–1941), the man who had received Albert Einstein in 1923 (Glick, 2005: 123–32); years later he vilified 'anti-Spanish internationalism' and even served as president of the commission to purge the university teaching staff during the Civil War. Subsequently, of course, Rocasolano was appointed vice president of the CSIC and in 1946, five years after his death, the CSIC Institute of Physics and Chemistry was named after him (Otero Carvajal, 2014: 102).

Chapter 16

Reading Science in Franco's Spain: The Early Years

In the words of Enrique Moles, writing in November 1910, 'without having access to the collections of specialist journals, one cannot try to produce original works' (Moles, 1911: 84–5). The old conviction about the need to consult scientific journals was very much alive in post-Civil War Spain and during or even shortly after the Second World War, when steady contact with the outside world became as complicated as it was vital for the scientists at the Spanish National Research Council. At that time, in the mid-1940s, scientific journals and books from Germany, France, England and America were mainly received via the CSIC International Exchange Section or through booksellers such as Rudolf Kadner – who managed twelve subscriptions to German journals, some of which encountered mishaps during the mailing process[1] – José Bosch and the Spanish Booksellers General Partnership.[2] These acquisitions comprised a good part of the chemistry and physics library collections, which, as in the other CSIC institutes, were subject to a restricted lending service:

> only for the gentlemen who work in the Institutes, so they can read the books in the laboratories outside Library hours . . . The books cannot be lent to remove them from the Institutes. Readers who have no connection to the Institute who wish to read in the Library must have a library card signed by the Director or one of the Secretaries.[3]

These reading conditions – and constraints – are reminders that libraries also act as regulated and normativised spaces, as a means of social relations and organisation that, while making reading possible, require that it be done under a specific set of conditions that allow the reader to use and enjoy the collections and the lending system, as long as they meet the bureaucratic requirements to do so. Structured, regulated and normativised, the library space in the CSIC Alonso Barba and Alonso de Santa Cruz Institutes also provides information about the what, how and who of reading during the dictatorship, always according to the legislation of the time, which separated people and restricted some reading to recognised expert readers (García Naharro, 2020a). Here, both the definition of 'legitimate library users' and the regulation of access to the collections often resulted in the consultation of specialist bibliographies being negotiated through personal contacts or informal channels –conversations, personal letters, requests to third parties, and so forth – thus demonstrating that in Spain at that time, an interest in consulting a foreign scientific article or book was not unique to the CSIC researchers:

> To be able to more efficiently continue tests that we are currently conducting to obtain new products, it would be very useful if we could consult the recent chemistry bibliography, and for that reason, we kindly ask that you allow our employee, the Graduate in Chemical Sciences, Mr Carlos Velasco Corral, access to the Library in the centre that you direct.[4]

With these words, written in September 1944, the managing director of the soap factory La Madrileña S.A. wrote to José Casares Gil, the director of the CSIC Institute of Physics and Chemistry, providing him, in turn, with the necessary credentials to win his trust and the requested permission. This same strategy was used by Fernando Martín in June 1947, when he contacted Julio Palacios to ask him about a study on electrostatic machines published in 1930 in *Comptes Rendus de l'Acedemie des Sciences*, one of the journals received by the Alonso de Santa Cruz Institute:

> I kindly request that you inform me if you could lend me, for one or two days, the cited volume. If so, I will send a trusted messenger to pick it up and I will return it the same way. As for myself, if the official centre to which I belong and of which I am the director does not provide guarantee, you can request reports from José Baltá Elías, who encouraged me in the search for this work, cited in one of his recent studies.[5]

Although being referred by José Baltá Elías (1893–1973), a chair professor in electromagnetism at the Universidad de Madrid and head of the electricity section at the Alonso de Santa Cruz Institute itself, was a complete guarantee, the regulations were clear: 'Lending books from this Library is categorically prohibited'.[6] Thus, the library collection was established as being for the exclusive use of the council staff, or other duly authorised individuals, with strict restrictions on outside lending, including between staff at other CSIC institutes, and with the possibility of monetary sanctions in the case of non-compliance with the lending regulations. In April 1948, Manuel Colomina wrote to Raimundo Menéndez:

> I am sorry to inform you that, due to the serious infringement of the regulation regarding the lending of a book from this Library, it has been decided to sanction you with a fine of 50 pesetas and a prohibition on taking books out of this Library for a period of three months. Non-compliance with the financial sanction will be accompanied by the prohibition of access to the Library. All the books that you currently have out on loan must be returned as soon as possible.[7]

This measure was justified by the frequency with which books disappeared from the CSIC libraries, which led some institutes to take 'the radical decision to stop book lending and to provide every opportunity to use the books in the Library to the Institute's collaborating staff'.[8] Although this degree of control was not the norm, strict controls were stipulated over who could borrow books and who could not: 'We will only loan books from the

Council Libraries to collaborators and fellows at those Institutes, always upon the request of the Director, Secretary or any Section Head of the corresponding Institute, as in this way the Institute will be able to respond to the loan request.'[9]

Under these conditions, the movements of readers at the Alonso Barba and Alonso de Santa Cruz Institutes varied significantly during the mid-1940s. As the library director, María Fernández Vallespín, recalled, 'Sometimes the library is full, and other days only 10 to 12 readers pass through it'. However, according to its files, in March 1945, the library lent out 314 books and 172 journal issues out of the 1,413 books and nearly 4,000 journals it had at the time.[10] Sadly, these numbers lack faces and names. Who were these readers? What printed products did they read? However, and despite the scarcity of the sources, there is evidence that, for instance, one of the salient CSIC researchers, José María Otero Navascués, consulted *Física Moderna* by Gaetano Castelfranchi, a member of the Consiglio Nazionale delle Ricerche in Italy;[11] and that Julio Palacios was interested in the works of renowned authors such as Sir Arthur Eddington and his *The Mathematical Theory of Relativity* and *The Internal Constitution of the Stars*, Max Planck's *Vorlesungen über die Theorie der Wärmestrahlung* and Edmund C. Stoner's *Magnetism and Matter*. A Mr Beltrán checked out works like *Matiere et energie* by Victor Henri, *Microenergetic* by P. Bricout, *Molkülstruktur* by Herbert Arthur Stuart, and *Infra-red und Raman Spectra* by C. B. Sutherland, while a Mr Bartolomé was interested in works like *Leçons sur la conductibilité des électrolytes* by Eugène Darmois. Incidentally, all these copies came from the confiscated personal library of Blas Cabrera, as did *La mécanique des phénomènes fondée sur les analogies* by Mihailo Petrowitch, a book that González del Valle asked to consult before it was returned to Cabrera's heirs.[12]

Thanks to Luis Solana, a doctor in sciences, military aviator and professor at the *Escuela de Cuatro Vientos* air academy, it is known that in mid-1946, all the history books belonging to Enrique Moles's library were in the hands of Julio Guzmán, that the Landolt-Börnstein tables (Numbers 374 to 379) were held in the office of Octavio Rafael Foz Gazulla in the Faculty of

Sciences in Madrid and that many of the electrochemical works missing from the library were probably in the hands of Antonio Rius Miró.[13] By that time, Julio Palacios had already acquired the book by Louis de Broglie, *De la mecanique ondulatoire a la théorie du Noyau* (I and II), which he had bought in November 1945 through Charles Ross (66 Alfonso XII Street Madrid) for 250 pesetas, with the costs incurred, as always, by the Alonso de Santa Cruz Institute.[14]

Table IV.4. Books requested by Julio Palacios for the Alonso Barba and Alonso de Santa Cruz Institutes Library (1946–7)[15]

Author	Title	Publisher	Year of publication
American Institute of Physics	*Temperature*	Reinhold	1941
American Society of Refrigerating Engineers	*Refrigerating Data Book*	ASRE	1939
Barnard, Ellenwood and Hirschfeld	*Heat-Power Engineering*	Wiley	1926–30
Bošnjaković	*Technische Thermodynamik*	Steinkopf	1935
Church	*Steam Turbines*	McGraw-Hill	1935
Epstein	*Thermodynamics*	Wiley	1937
Fermi	*Thermodynamics*	Prentice Hall	1937
Gibbs	*Collected Works*	Longmans	1931
Hoare	*Thermodynamics*	Longmans	1936
Kearton	*Steam Turbine Theory and Practice*	Pitman	1931
Keenan and Keyes	*Thermodynamic Properties of Steam*	Wiley	1936

Kiefer and Stuart	*Principles of Engineering Thermodynamics*	Wiley	1930
Lewis and Randall	*Thermodynamics*	McGraw-Hill	1923
MacDougall	*Thermodynamics of Chemistry*	Wiley	1939
MacIntire	*Refrigeration Engineering*	Wiley	1940
Noyes and Sherrill	*Chemical Principles*	Macmillan	1938
Stodola-Lowenstein	*Steam and Gas Turbines*	McGraw-Hill	1927
Taylor and Taylor	*The Internal Combustion Engine*	International	1938
Walker, Lewis, McAdams and Gilliland	*Principles of Chemical Engineering*	McGraw-Hill	1936
Warburton	*Master's Thesis*	MIT	1938
Weber	*Thermodynamics for Chemical Engineers*	Wiley	1939
Zemansky	*Heat and Thermodynamics*	McGraw-Hill	1937

Months later, while other CSIC researchers were trying to complete their private collections of the annual reports of the former Board for Advanced Studies and Scientific Research (JAE),[16] Otero Navascués was insisting on acquiring the *Journal of Optical Society* for himself. However, the director of the council libraries told him that it would be preferable for the collection to remain in the institute library, 'because it would be a shame to break it up and create little libraries'.[17] To encourage compliance, in October 1947 it was stipulated that all requests for books for the Antonio de Gregorio Rocasolano Institute for Physical Chemistry library had to go through the director, then Manuel Colomina

Barberá.[18] However, sporadic cases of individual requests continued to occur. For instance, around that time, the secretary of the Rocasolano Institute for Physical Chemistry wrote a letter to the Heffer & Sons publishing house on the matter:

> To my great astonishment, given that in my letter of 14 July 1948, I told you not to send any more books or journals related to the order made by Mr Belenguer in August 1947, this Institute received, in the name of Mr Belenguer, the following books: Berkman, *Catalysis; Annual Reports on the Progress of the Chemistry*. I would appreciate it if you could tell me who ordered these books and that henceforth no more books be sent in the name of Mr Belenguer, as a member of this Institute, for which, due to acquisition difficulties, I cannot be responsible. To prevent further concern on your part and ours, I ask that you only send those books ordered for us by Dr J. M. Gamboa, after payment of the sum, and those that I order as Secretary of this Institute, which I will only do when I know that it will be possible to pay you shortly thereafter.[19]

From that time on, Colomina and Gamboa centralised purchases for their library, which continued to be supplied by the CSIC International Exchange Section and increasingly by purchases from publishing houses and distributors, exporters and importers, including the Sociedad General Española de Librería (9 Evaristo San Miguel, Madrid), Distribuidora Europa (70 Avenida de José Antonio, Madrid), Editorial Pueyo S.L. (5 Tetuán, Madrid), Afrodisio Aguado S.A. (124 Ayala, Madrid), Manuel Aguilar (22 Serrano, Madrid), Editorial Herder (25 Balmes, Barcelona), Editorial Labor S.A., Librería General de Zaragoza, Enrique Villain (commercial agent, 16 Avenida Reina Victoria, Madrid), Ediciones Nueva Época S. A. (17 Serrano, Madrid), Librería Paraninfo (65 Meléndez Valdés, Madrid), Librería J. Villegas (33 Preciados, Madrid), Librería Anaquel, Peter Petsch Arnold (Russian journals, 4 Hilarión Eslava, Madrid), Dossat and the CSIC's own Medinaceli Scientific Library, among others.[20]

Although many of the names of these publishers and booksellers were from the commercial sphere, they nonetheless played a role in importing or subscribing to foreign scientific journals, thus gaining access to so-called 'higher science' circles, even if their usual clients and the potential consumers for their other editorial products did not inhabit that space (García Naharro, 2015b and 2019). Beginning in the mid-1950s, it became much more common – in line with the prevailing business model amongst international journals and the major labels – for the CSIC to buy directly from foreign research centres, booksellers and large publishing corporations: Lange, Maxwell & Springer (New York), McGraw-Hill (New York), Centre National de la Recherche Scientifique, Electrochemical Society, the United States Government Printing Office, William & Wilkins, Comptes Rendus de l'Académie des Sciences, Simpkins, Pergamon, and the like.[21] This was a time when the scarcity of the post-Civil War years became a thing of the past, and the dictatorship embarked on its gradual, strategic process of opening up and making domestic readjustments.

However, even in the early 1950s, the director of the CSIC libraries continued to recognise the precarious state of their catalogues and the organisational system in the different council libraries, asserting, moreover, that 'its most serious problem' continued to be the acquisition of collections.[22] Accordingly, in 1953, Javier Lasso de la Vega, the director of the library at the Universidad de Madrid, member of the National Institute for Labour Rationalization, the International Organization for Standardization (ISO) and the International Committee for Scientific Management (CIOS), also acknowledged the need to implement library planning in Spain, even going so far as to say that Spanish researchers continued to lack access to a large number of specialist journals, unable to avail themselves of them due to the difficulties characteristic of, he said, 'a country as poor in resources as Spain'.[23]

When the CSIC took stock of the situation between 1952 and 1954 in the second volume of the corresponding *Memoria de la Secretaría General* – which was not published until 1958 – the council expressly recognised the worrisome circumstances surrounding a large number of council libraries:

> We have very closely read the Annual Reports sent from the directors of the libraries in the Council's Institutes. Very few do not lament the impossibility encountered during these last few years of acquiring books and journals; many Institutes today do not have any form of income beyond the procedure of exchanging their own publications, and bearing in mind that the best foreign journals do not engage in exchanges due to the fundamental entrepreneurial nature of the respective publishing houses, it is easy to understand how many of the libraries' needs remain unmet. How, little by little, the most interesting collections are becoming incomplete, and even the most prestigious men of science must complain in the press, as recently occurred in reference to the Botanical Garden, that its extraordinarily rich library, containing unique copies in the scientific world, is lacking in modernity regarding its collections. . . . These opinions . . . which are frankly critical about the poverty of the current collections in a specific library, can perhaps be applied to many more. The libraries of the Council Institutes today generally suffer much hardship. Many of their collections continue to be incomplete and they cannot subscribe to journals that are considered highly useful. The lack of means in the Institutes is inevitably reflected in their libraries, of which it can be said that their acquisitions have been reduced by half over the last three years.[24]

'The starting point must be the obvious reality that we are a poor country', admitted José Ibáñez Martín in a letter written in December 1961 to the then Minister of Finance, Mariano Navarro Rubio. After congratulating him for his work and for the budget surplus of the previous fiscal year, Ibáñez Martín asked Navarro Rubio to create a special research service to relieve 'the pressing need to more energetically protect research':

> Look into this interesting topic. Talk to the Caudillo about this matter, and with his protection and his stimulus, believe that in the field of research, Spain can be given something at least reaching the level of small countries such as Switzerland, Belgium and Holland, which are comparatively insignificant compared to us in terms of land and population, but today have technology . . . that is

> comparable with the most important countries in the world. This is the only path that can make Spain into a great country.[25]

Far from public rhetoric and, perhaps, protected by the trust and confidence offered by a private letter, Ibáñez Martín's words brim with the significance of a public question that, paradoxically, continues to be as valid today as it was back then. Perhaps for that reason, his words, so unprecedented and disconcerting, resound with familiarity today: 'As long as Spain does not support scientific research, both basic and applied, it will continue to be a country without a future.'[26]

Conclusion

To fully make sense of the copious collection of ideas and data scattered throughout the pages of this book, it is necessary to organise the main arguments developed in the four parts that comprise the text. The linkage and arrangement of the structure construed from 'Words', 'Actors', 'Printed Objects' and 'Places' in the book correspond not only to a methodological dimension, but also to a series of considerations about the subject itself. Epistemological questions, given institutional form from the scientific field under a dictatorship and demarcated by spaces and printed products associated with the 'expert' as legitimised by the regime, are analysed here through the fragmentary study of specific places and social actors, political discourses and scientific publications within the spatiotemporal framework of Franco's Spain. The specific focus is on the period between 1939 and 1966, a concrete moment in the history of the Spanish National Research Council, from its foundation to its restructuration by decree on 1 December 1966. That decree grouped the original eight boards established in the founding law into four, with institutes or centres of equal status and similar concentrations being converted into national institutes. This reform, combined with term limits, turnovers and a system to elect directors, sought to invigorate the institution and adapt it 'to the demands of its natural growth and the scientific movement itself'.[1]

At the same time, this particular time interval also harkens back to a specific period in the Franco dictatorship, when the editorial output in Spain was constrained by wartime censorship and the official privileges prevailing over the private initiative of a few editors who, notwithstanding, found in the politics of influence their primary tool for survival. Indeed, to a large extent, part of what makes this book both innovative and relevant is its central claim to use printed culture as the starting point to view texts – in the broadest sense of the word – and the social functions of the written word, its forms and formats, relative to the configuration of power relations, both in the scientific field and the field of scientific publishing during the Franco dictatorship. Although these two spheres are closely related, they have, paradoxically, received little academic attention using a consolidated perspective like the one developed in this book, whose main objective is to reconcile the contributions from the history of science with materials of a culturally considered social history.

Studies that have analysed science in Spain and, specifically, those that have focused on the Spanish National Research Council and its associated centres have traditionally done so from the point of view of its institutional history, and there can be no doubt that these works have been of great benefit when trying to understand the genesis and development of science during the Franco period. However, they have not always taken into account (or at least not in such detail) the facet of the CSIC as, in the words of José María Albareda writing to publisher Gustavo Gili in April 1944, 'a buyer of books, a publisher of diverse publications, or the recent initiative to make it responsible for reviewing the scientific works that are up for publication' (García Naharro, 2019: 27). This aspect of the Council as the official scientific institution responsible for approving scientific and technical works in the world of Spanish publishing, along with its role in Spain as the authority able to consecrate science – in close collaboration with the political arena – demonstrates how, for certain names, editorial products and spheres of action, merely being associated with the CSIC gave them the capacity of agency to influence Spanish publishing and science through the material or symbolic benefits they were qualified to distribute (Bourdieu, 1995: 82).

'As you can see, he operates in circles to which it would behove us to have access.' Using these forthright words, the technical advisor at the Gustavo Gili publishing house, Pedro Reverté, referred to Salvador Terol Alonso, a doctor in chemical sciences and director of the section of the same name in the Laboratories of the General Staff of the Navy, as well as head of the luminescence section of the CSIC's Daza de Valdés Optics Institute (García Naharro, 2019: 90–1). In October 1956, that same publishing house also wrote to Eduardo Torroja, director of the Technical Institute for Construction and Cement at the CSIC, to see if he could augment the Spanish-language publication of a work by the Italian engineer Pier Luigi Nervi with a text of his own, because of 'the deserved prestige that you enjoy, not only in our country but even abroad' (García Naharro, 2022: 242–3). These and other acts of politically and socially recognised credit – from the awards, viewed as signs of distinction with economic dividends, to the laudatory reviews and the articles co-authored by young researchers and figures acclaimed by the scientific and academic institutions of the Franco period – bear witness, as demonstrated time and again throughout this book, to how the official scientific institutions in Franco's Spain held a monopoly over the definition of science and what is scientific or, in other words, the power to determine what science was – and was not – in Spain during that period.

However, reconstructing the social and historical conditions of the reality that shaped the entire context of producing, circulating and receiving printed scientific products in Franco's Spain requires applying a cross-sectional approach, not only to the categories that characterise the scientific field, but also to those controlled from the field of political power. This entails paying attention to the official – and even supposedly ideal – criteria of scientificity, as well as the ideological elements involved in the processes of constructing and demarcating what is 'scientific' in the context of a dictatorship. To that end, this book has availed itself primarily – but not only – of an analysis of the speeches given by José Ibáñez Martín as the authorised voice of the Franco regime and salient figure of the CSIC in the first part of the book, examining the symbolic work of representing a specific scientific identity: the official identity. These speeches and the discourse conveyed in them provide a fitting textual and discursive space to apply a semiotic perspective to the arguments set forth and the displacements made through acts of citing that brought together a variety of references and produced effects of reality in texts that, however, also played a part in altering that very reality.

Thus, by underscoring both what underlay these speeches and what they expressed, this analysis demonstrates how the so-called official image of science mapped out in the texts was gradually tempered and adapted to the events of the times. Over the years, the authoritarian post-war science, represented as distinctly national, unitary and neothomistic in inspiration, transformed into an image of a more modern and allegedly competitive science, defined by the creative adaptation of foreign knowledge in a Spain decked out at the time for the '25 Years of Peace' celebration. However – and herein, perhaps, lies the innovative aspect of this work – the analysis of these speech acts transcribed onto paper show how the official discourse led to a diverse range of categories and properties that were closely related to specific actors, objects and places that ultimately set the contours of a sought-after legitimate science – legitimated by the regime – while establishing relationships of belonging and drawing boundaries between groups and actors represented in the social space.

Echoing the formulation put forth by Roger Chartier, the concept of representation suggests a view of the social world or the exercise of power as a system of relationships (Chartier, 2000: 85). Similarly, the analysis of the discursive and material devices used to present the official representation of science during the Franco period has provided an opportunity to delve – again, cross-sectionally and throughout the book – into the mechanisms and forms of inculcating particular conventions and the subsequent production (or lack thereof) of effects. To that end, in addition to the pertinent critical review of documents, it was deemed necessary to analyse the social status of the documents themselves. This applied not only to the most obvious case of the transcribed speeches of Ibáñez Martín, originally delivered aloud to a particular audience and in a specific space of enunciation, but also to the reports, catalogues, journals, letters and other primary and secondary sources that have appeared over the course of the book, providing documentary support to the assertions. In their own historicity, these products recognise and also materially experience the weight and agency of the official image of science in Franco's Spain.

When compared and contrasted, these documents reveal past perceptions and realities, through the texts thought to recognise, describe, prescribe or proscribe them, all, of course, without being carried away by an illusion of absolute political efficiency or confusing, in practice, the different logics that govern the production of discourses and the regulation of behaviours, respectively (Chartier, 2000: 84). It is, precisely, the irreducibility of all practice to discursive strategies that has spawned many of the epistemological concerns that guide the study of the written and printed products that fill the pages and lead the sections of this book. Without doubt, the CSIC's scientific and technical publications deserve special mention. Intentionally given centre stage in the book, they formed a foundation to introduce other important figures like the scientist (as actor and author), in turn used to get to the bottom of the conventions that operated behind their recognition as legitimised experts under Franco.

This book, then, has examined the significance of the physical form of texts turned into a material, textual and discursive space where, in contrast to usual practices, it is possible to confirm not only the potentialities – as an instrument of power – of the official discourse and its influence when conceiving, for example, scientific journals, but also the sphere and very practice of science, as a prime locus for the male/subject/knower (*vir modestus*) of the CSIC. These male scientists, seen and presented as reliable authorities with 'purity of blood' (*limpieza de sangre*), monopolised the positions of responsibility and dominated the pages of the CSIC's scientific publications although, and as analysed in the second part of this book, a number of women, using their proper names – and, for the most, complying with shared profiles and patterns – also figured out a way to find a place for themselves.

Indeed, the proper name, a fundamental category and the most precious property when the name is renowned (Bourdieu, 2008: 19), receives a significant amount of attention in these pages.

Undeniably, if credit and authority in science are associated with diverse names, one of them – perhaps the most powerful – is that of the scientific author, a figure constructed here from their inscription in texts and the way those texts are ultimately used in practice when inserted into the printed format. Accordingly, this book has tried to demystify these figures in the interest of the relationships in which scientific authors – male or female – were constituted as such. This has unveiled not only the essentially relational character of the scientific credit behind the names of the scientists, but also that of official science itself under the Franco regime.

The resulting institutional science was promoted and sustained by the massive editorial output of the CSIC, which never lacked the paper it required. For example, while the publishing house Labor had to work without the necessary raw materials to finish the publication of technical works like *Tecnología Mecánica* by José Serrat y Bonastre and G. D. Jerie's *Elementos de Máquinas* in 1940,[2] in December of that same year, Alfredo Sánchez Bella wrote to the various bodies in the council to find out what 'type of paper' they wanted to use, whether 'the quota of paper assigned to their periodical publications was sufficient' and, if it was not, 'the amount by which the quantity should be increased'.[3]

These CSIC scientific periodicals clearly take pride of place in the third section of this book. They have been used to illustrate the entire endogamic system, made up of works primarily written in Spanish, which operated as an active tool in the formation and maintenance of the Spanish scientific community under the dictatorship. Moreover, the Franco regime wished to use these official scientific journals to project the image abroad of a robust science that did not always exist. As Alberto Sols wrote to his family in February 1952, '99 per cent of Spanish scientific output is completely lost in the cemetery of journals that nobody reads, firstly because they do not find them interesting, and secondly because of the linguistic demands' (Santesmases Navarro de Palencia, 1998: 69).

The scant international presence of Spanish scientific publications (Malet, 2008), however, contrasted with the high number of

CSIC publications that received institutional coverage, dissemination and financing. As this book has shown, these publications were subject to the rules, resources and technical means available to the regime at a time when the government always supported state production to the detriment of private publishing initiatives (García Naharro, 2015a and 2019). Leaving aside traditional analyses and evaluations of the contents of scientific periodicals, the third section of the book presents a detailed study of the material dimension of the journals produced by the CSIC's Juan de la Cierva Board. Consistent with the book's overall desire to demarcate science with its narrative, the aim of this approach is to puzzle out what exactly turned these journals into sorts of 'boundary objects' (Star and Griesemer, 1989; Star, 2010), shaped by editorial projects, material formats and spaces of circulation. They are understood as printed products whose pages unite a set of scientists legitimised by the regime with a specific community of actors attached in some way to the scientific field, also showing how the CSIC journals became prime places to promote specialist publishers, industries and companies that sought visibility on the pages of publications that, from their perspective, were printed vehicles that circulated in attractive national, and even international, spaces.

In this respect, along with the study of the processes of constructing scientific authority previously examined in the book, this analysis of the journals, understood as printed products combined with their associated value and the characteristics and profiles of their consumers, opens up a whole range of possibilities related to the differentiated meanings and uses that journals can acquire according to the social areas where they circulate and the interests and abilities of their readers, users or mere consumers. All this, in turn, means that the place where the scientific journals could usually be found in Franco's Spain merits some degree of attention: thanks to their subjects and the high cost of acquiring a complete collection, national and international scientific journals usually found their niche in the libraries of the institutes of higher science, such as those found at the Spanish National Research Council.

As such, these libraries had an inherent order and specific configuration, although their normativity and apparent stability were altered by the intervention of readers and users who, through their practical relationship with the scientific output available there, converted the library 'space' into a practised 'place' (De Certeau, 2000: 129–31). To this end, the fourth part of the book is dedicated to the study of the archives of the Alonso Barba and Alonso de Santa Cruz Institutes Library, and the uses and relational modes of the CSIC research community with these printed products. However, while libraries in Spain – both public and private – present a choice space in which to try to connect with the readers of the past (García Naharro, 2020a: 105–8), it is rather unusual to study the libraries of scientific institutions, and even more so from the perspective applied here, which seeks to reconstruct the material conditions for the reception of knowledge.

From this perspective – and taking into account the regulated and normativised space of the library – this work has shown that, even during the heightened international isolation of the Franco regime, the CSIC scientists had access to foreign scientific content. Despite numerous obstacles and an endemic lack of budgetary funds, the council found an effective tool to remedy these shortcomings in the exchange of publications and, even, the confiscation of library collections. Moreover, as demonstrated by the titles taken from Blas Cabrera and Enrique Moles, these books and journals highlight the aporias of a dictatorial regime that simultaneously vilified their legitimate owners and offered the CSIC researchers the bibliographic heritage compiled by the two scientists during their lives, which was seen as having value. This worth was also unequivocally attested to by the interest shown on the part of these researchers in consulting the confiscated books and journals that, paradoxically, thus became printed products that were somehow able to circumvent the authority of the Franco regime and, once again, disarm the rhetoric of its official discourse. This discourse itself has, in some way or another, been a constant companion throughout the pages of a book that ends here.

Epilogue

The Resilience of Spanish Science in the Face of Loss and Injury

> The recovery of historical memory and the recognition of the Spanish scientists, both men and women, who were persecuted during the Franco dictatorship is a matter of justice for the CSIC. Moreover, it constitutes an essential step towards fully understanding Spain's scientific legacy and recognizing the value of science for the progress of society. It also amounts to a recognition of democracy as a form of government that guarantees liberty, development and opportunities.[1]

With these words, the president of the Spanish National Research Council, Eloísa del Pino, summarised the reasons for the publication of a 2024 report on how the CSIC has been handling democratic memory and has adapted to the official measures related to the recovery of memory promoted in Spain by the 2007 Law of Historical Memory and the 2022 Law of Democratic Memory. This document, which delves into the deep impression left on the institution by the Franco regime, was drafted by a working group assembled in 2023 by the CSIC president and made up of various Spanish researchers from the council: Ana Romero de Pablos (Institute of Philosophy), María Jesús Santesmases (Institute of Philosophy), Leoncio López-Ocón (Institute of History) and Francisco Ferrándiz (Institute of Language, Literature and Anthropology), all experts on the subject.

Their efforts consisted of working out measures aimed at complying with the provisions in the Law of Democratic Memory and thus encouraging reflection in the CSIC on its past with regard to the Franco dictatorship. The elements subject to consideration under the law included, for example, the names of some council institutes and centres that referred to individuals associated with the dictatorship and, particularly, its repressive policies,[2] such as the [Antonio de Gregorio] Rocasolano Institute for Physical Chemistry. Although the working group proposed renaming the institutes and centres according to their discipline alone as a general criterion, the members were open to using the name of a female researcher agreed on by the staff of this centre, thus incorporating the dimension of gender into the institution's memory policies. According to the 2024 report, it was actually the CSIC president herself who proposed replacing the surname Rocasolano with the names of the scientists Blas Cabrera or Enrique Moles, if the board and staff of the Institute for Physical Chemistry deemed it appropriate, and in spring 2023, the two bodies unanimously decided to rename their research centre the Blas Cabrera Institute for Physical Chemistry (IQF).[3] The commemorative plaque located in the vestibule of the institute explains that 'the new name pays homage to the memory of Blas Cabrera y Felipe, a brilliant scientist and one of the fathers of Spanish physics, and the first director of the National Institute of Physics and Chemistry, located in this very building and inaugurated on 6 February 1932'.[4]

Another feature potentially affected by the law was located in the main CSIC building, in the north wing of the first-floor hallway, right next to the entrance to the president's office: the portrait gallery of the presidents of the Board for Advanced Studies and Scientific Research (JAE) and the CSIC, inaugurated in 2021. In the opinion of the working group, this pictorial, material and virtual tour (also available on the CSIC website) did not reflect the historical fracture between the two institutions. To re-signify this gallery, the experts recommended keeping the portraits of all the presidents, including those appointed during the dictatorship, but converting it into a space of democratic memory for the council, 'to remember the origin of the CSIC in the Franco regime and the involvement of some of its presidents, especially the first one, José Ibáñez Martín, in the dictatorship's repressive policies'.[5] However, in accordance with the criteria established in the 2022 Law of Democratic Memory, on 12 April 2023, the president of the council arranged to remove the image of Ibáñez Martín – painted 'against the backdrop of the Spanish flag and national coat of arms under Franco' – as representative of the repressive history of the CSIC. His portrait was later moved to the council's Historical Properties Archive, as explained by the text on the plaque that currently fills the space where the portrait of Ibáñez Martín once hung.[6]

This was not the only portrait of the honorary president of the CSIC[7] to be removed in the spring of 2023. In the southern wing of the main hallway on the first floor of the same building, another portrait of José Ibáñez Martín, painted by Alfonso Hernández Noda, hung alongside that of the CSIC's first secretary-general, José María Albareda, by Luis Mosquera Gómez. According to the report, 'given the involvement of the representatives in the repressive actions of the CSIC against the scientific personnel who were loyal to the Republican regime', the working group recommended taking down both portraits and placing them in the archive, leaving in their stead two plaques that explain to the visitor that they were removed in April 2023 'in compliance with the provisions in Article 34 of Law 20/2022'.[8] On 29 May of that same year, a small bust of José María Albareda sculpted by Jacinto Higueras in 1968 was also removed and replaced by a plaque. It had originally stood where the large bronze sculptures of the Nobel Prize winners Ramón y Cajal and Severo Ochoa sculpted by Víctor Ochoa (the great-nephew of the Nobel winner) now stand, having being inaugurated in 2007 to coincide with the centennial of the creation of the JAE.

Article 35 of Law 20/2022 of Democratic Memory also applied to the CSIC's signature books (*libros de honor*) 'because they contain the signatures of Franco and other prominent figures from the Franco regime'. The working group recommended that these books continue to be used as a 'written record of the institutional memory' of the council, but that explanatory inserts be included that make it clear that the CSIC 'responsibly recognized its past and lays claim to its memorial future'.[9] In this respect, the CSIC president also decided to perform an act of institutional reparation on 6 July 2023 'in homage to and memory of all the scientists at the JAE who were persecuted during the Spanish Civil War and the Franco Dictatorship'. This act began in the vestibule of the Blas Cabrera Institute for Physical Chemistry, followed by the inauguration of a plaque located opposite the façade of the main CSIC building that contains a QR code listing the names of all the persecuted scientists. The ceremony concluded in the new CSIC Democratic Memory Cloister in Madrid, a courtyard adjacent to the Church of the Holy Spirit (the former auditorium of the renowned *Residencia de Estudiantes*), with a commemorative plaque in memory of the JAE's former ownership of the structure.

'To recognize the CSIC as an institution that symbolizes Spain's ability to transform and overcome.' This was also the aim of the speech given by Eloísa del Pino to mark the eighty-fifth anniversary of the CSIC in the presence of Ministry of Science, Innovation and Universities Diana Morant, Secretary of State for Science, Innovation and Universities Juan Cruz Cigudosa, Secretary-General for Research Eva Ortega Paíno and several authorities and former council presidents: to demonstrate that the CSIC of the twenty-first century 'is a legitimate and effective institution in democratic terms . . . that, therefore, looks to the future without forgetting its roots'. During this act, held on 17 December 2024, the Spanish biophysicist Eva Nogales was presented with the CSIC Extraordinary Medal for Scientific Merit for her work in the field of the visualisation of macromolecular functions.[10] The medallion (formerly known as the CSIC Gold Medal) continues to feature the council emblem, the *Arbor Scientiae*, although it has been stripped of the banners and descriptors that once adorned it. However, the old version of this Tree of Science continues to dominate in places, such as the stained-glass windows in the central staircase and east wing of the first floor of the CSIC's main building. This material symbol, which has been a constant throughout this book as a tangible reminder of the iconographic programme of the Franco regime, continues to be quite well-established, in both the aesthetics of the building and in the collective imaginary of the institution. 'Preserving it is beyond question', asserts the CSIC report on how to manage democratic memory at the council, *El tratamiento de la memoria democrática en el CSIC*.[11] This explains why the stained-glass windows with the *Arbor Scientiae* have not been removed from the staircase in the main building. However, they are now accompanied by a plaque that bears the following inscription:

> The symbol of the CSIC, the *Arbor Scientiae*, in this representation, places theology as the primary trunk from which the rest branch off. This conception of National-Catholic science provided the framework for the act of purging scientists during the Franco period and does not represent the Institution.[12]

Select Bibliography

Albareda Herrera, José Mª, *Consideraciones sobre la investigación científica* (Madrid: CSIC, 1951).

Aróstegui Sánchez, Julio (coord.), *Franco: la represión como sistema* (Barcelona: Flor del Viento, 2012).

Ballestero Aguilar, Alfonso, *Juan Antonio Suanzes, 1891–1977: La política industrial de la posguerra* (Madrid: LID Editorial Empresarial, 1993).

Barona Vilar, Josep Lluís, *El exilio científico republicano* (Valencia: Publicacions de la Universitat de València, 2010).

Biagioli, Mario (ed.), *The Science Studies Reader* (London and New York: Routledge, 1999).

Biagioli, Mario, and Peter Galison (eds), *Scientific Authorship: Credit and Intellectual Property in Science* (London: Routledge, 2003).

Bourdieu, P., 'The Specificity of the Scientific Field and the Social Contradictions of the Progress of Reason', *Social Science Information,* 14/6 (1975), 19–47.

— *La dominación masculina* (Barcelona: Anagrama, 2000).

— *El oficio de científico* (Barcelona: Anagrama, 2003).

—*Intelectuales, política y poder* (Barcelona: Anagrama, 2012).

Calvo Calvo, Luis (ed.), *El CSIC en Cataluña (1942–2012): Siete décadas de investigación científica* (Madrid: CSIC, 2012).

Camprubí, Lino, *Engineers and the Making of the Francoist Regime* (Cambridge MA: MIT Press, 2014).

Camprubí, Lino, Xavier Roqué and Francisco Sáez de Adana (eds), *De la Guerra Fría al calentamiento global. Estados Unidos, España y el nuevo orden científico mundial* (Madrid: Los Libros de la Catarata, 2018).

Casanova Ruiz, Julián, *La iglesia de Franco* (Madrid: Temas de Hoy, 2001).

Claret Miranda, Jaume, *El atroz desmoche: La destrucción de la Universidad española por el franquismo, 1936–1945* (Barcelona: Crítica, 2006).

Csiszar, Alex, *The Scientific Journal: Authorship and Politics in the Nineteenth Century* (Chicago IL and London: University of Chicago Press, 2018).

Delgado Gómez-Escalonilla, Lorenzo, and Santiago López García (eds), *Ciencia en transición. El lastre franquista ante el reto de la modernización* (Madrid: Sílex, 2019).

Fernández Terán, Rosario E., *El profesorado del 'Instituto Nacional de Física y Química' ante la Guerra Civil, el proceso de depuración y el drama del exilio* (Madrid: UCM, 2014).

Formentín Ibáñez, Justo, Alfonso V. Carrascosa Santiago and Esther Rodríguez Fraile, *José Ibáñez Martín y la ciencia española: el Consejo Superior de Investigaciones Científicas* (Madrid: CEU, 2015).

Foucault, Michel, *El orden del discurso* (Buenos Aires: Tusquets, 1992).

Galison, Peter, and David J. Stump (eds), *The Disunity of Science: Boundaries, Contexts and Power* (Standford CA: Standford University Press, 1996).

García Naharro, F., 'Reshaping Scientific Journals. Applied Science and its audience under Franco (1939–1966)', *Journal of Iberian and Latin American Studies*, 27/1 (2021), 115–32.

— 'Ciencia de papel. Privilegios y autoridad en España durante el primer franquismo (1939–1959)', *Historia Contemporánea*, 68 (2022), 233–62.

— *Palabras, actores, objetos y lugares. El CSIC y la demarcación de la ciencia en la España franquista (1939–1966)* (Madrid: CSIC, 2022).

Gieryn, T. F., 'Boundary-Work and the Demarcation of Science from Non-science: Strains and Interests in Professional Ideologies of Scientists', *American Sociological Review*, 48/6 (1983), 781–95.

Glick, Thomas F., *Einstein y los españoles: ciencia y sociedad en la España de entreguerras* (Madrid: CSIC, 2005).

Gómez, Amparo, Antonio Fco. Canales, and Brian Palmer (eds), *Science Policies and Twentieth-Century Dictatorships: Spain, Italy and Argentina* (Farnham: Ashgate, 2015).

González Blasco, Pedro, *El investigador científico en España* (Madrid: CSIC, 1980).

González Blasco, Pedro, José Jiménez Blanco and José Mª López Piñero, *Historia y sociología de la ciencia en España* (Madrid: Alianza, 1979).

González Cuevas, Pedro Carlos, *El pensamiento político de la derecha española en el siglo XX. De la crisis de la Restauración al Estado de partidos (1898–2000)* (Madrid: Tecnos, 2005).

González Ibáñez, Carlos, and Antonio Santamaría García (eds), *Física y química en la Colina de los Chopos. 75 años de investigación en el edificio Rockefeller del CSIC (1932–2007)* (Madrid: CSIC, 2009).

Gordin, Michael, *Scientific Babel: How Science Was Done Before and After Global English* (Chicago IL: University of Chicago Press, 2015).

Guston, David H., *Between Politics and Science: Assuring the Integrity and Productivity of Research* (Cambridge: Cambridge University Press, 2006).

Gutiérrez Ríos, Enrique, *José María Albareda: una época de la cultura española* (Madrid: CSIC, 1970).

Herran, Néstor, and Xavier Roqué (eds), *La Física en la Dictadura. Físicos, cultura y poder en España 1939–1975* (Bellaterrra: Universitat Autónoma de Barcelona, 2012).

Ibáñez-Martín Mellado, José A. (coord.), *José Ibáñez-Martín. En el Centenario de su nacimiento* (Zaragoza: Instituto Fernando El Católico (CSIC), 1998).

Janué i Miret, Mariciò, and Albert Presas i Puig (eds), *Science, Culture and National Identity in Francoist Spain, 1939–1959* (Cham: Palgrave Macmillan, 2021).

Jasanoff, Sheila (ed.), *States of Knowledge: The Co-production of Science and Social Order* (London: Tylor & Francis, 2004).

Juliá, Santos, *Historia de las dos Españas* (Madrid: Taurus, 2004).

Krige, John, *American Hegemony and the Postwar Reconstruction of Science in Europe* (Cambridge MA: MIT Press, 2006).

Krige, John, and Dominique Pestre (eds), *Science in the Twentieth Century* (Amsterdam: Harwood Academic Publishers, 1997).

LaFollete, Marcel C., *Making Science Our Own: Public Images of Science 1910–1955* (Chicago IL: University of Chicago Press, 1990).

Latour, Bruno, *Nunca fuimos modernos. Ensayo de antropología simétrica* (Buenos Aires: Siglo XXI, 2007).

Latour, Bruno, and Steve Woolgar, *Laboratory Life: The Construction of Scientific Facts* (Princeton NJ: Princeton University Press, 1986).

Laudan, L., 'The Demise of the Demarcation Problem', in R. Laudan (ed.), *The Demarcation Between Science and Pseudo-science* (Blacksburg VA: Virginia Tech, Center for the Study of Science in Society, 1983), pp. 7–35.

Law, John (ed.), *A Sociology of Monsters: Essays on Power, Technology and Domination* (London and New York: Routledge, 1991).

López García, Santiago, *Ciencia, tecnología e industria en España. Herencias institucionales y nueva política científica en la constitución del Patronato 'Juan de la Cierva' (1939–1945)* (Madrid: Fundación Empresa Pública, 1993).

— *Saber tecnológico en la política industrial del primer franquismo* (Madrid: UCM, 1994).

— *El patronato 'Juan de la Cierva', 1946–1960: entre la unidad de la ciencia y el interés nacional* (Madrid: Fundación Empresa Pública, 1995).

López Sánchez, José Mª, and Alba Fernández Gallego, *A imprenta y tírese: 80 años de la Editorial CSIC* (Madrid: CSIC, 2021).

Lora-Tamayo, Manuel, *Lo que yo he conocido. Recuerdos de un viejo catedrático que fue ministro* (Cádiz: Federico Joly y Cia., 1993).

Magallón Portolés, Carmen, *Pioneras españolas en las ciencias. Las mujeres del Instituto Nacional de Física y Química* (Madrid: CSIC, 2004).

Malet, A., 'José María Albareda (1902–1966) and the Formation of the Spanish Consejo Superior de Investigaciones Científicas', *Annals of Science*, 66/3 (2009), 307–40.

Nieto, Alejandro *et al.* (coords), *Apuntes para una política científica. Dos años de investigación en el CSIC: 1980–1982* (Madrid: CSIC, 1982).

Nieto-Galan, Agustí, *The Politics of Chemistry. Science and Power in Twentieth-Century Spain* (Cambridge: Cambridge University Press, 2019).

Nye, Mary Jo, *From Chemical Philosophy to Theoretical Chemistry: Dynamics of Matter and Dynamics of Disciplines, 1800–1950* (Berkeley CA: University of California Press, 1993).

Ortiz Gómez, Teresa, and Gloria Becerra Conde (coords), *Mujeres de ciencias: mujer, feminismo y ciencias naturales, experimentales y tecnológicas* (Granada: Universidad de Granada, 1996).

Otero Carvajal, Luis Enrique (dir.), *La Destrucción de la ciencia en España. Depuración Universitaria en el franquismo* (Madrid: UCM, 2006).

—— (dir.), *La Universidad nacionalcatólica. La reacción antimoderna* (Madrid: Universidad Carlos III de Madrid-Dykinson, 2014).

Otero Carvajal, Luis Enrique, and José María López Sánchez, *La lucha por la modernidad. Las ciencias naturales y la Junta para Ampliación de Estudios* (Madrid: CSIC-Amigos de la Residencia de Estudiantes, 2012).

Pérez Sedeño, Eulalia, and Paloma Alcalá Cortijo (coords), *Ciencia y género* (Madrid: Complutense, 2001).

Pimentel, Juan, *Fantasmas de la ciencia española* (Madrid: Marcial Pons, 2020).

Portero Rodríguez, Florentino, *Franco aislado. La cuestión española (1945–1950)* (Madrid: Aguilar, 1989).

Puig-Samper Mulero, Miguel Ángel (ed.), *Tiempos de investigación. JAE-CSIC, cien años de ciencia en España* (Madrid: CSIC, 2007).

Romero de Pablos, Ana, *Cabrera, Moles, Rey Pastor. La Europeización de la ciencia, un proyecto truncado* (Madrid: Nivola, 2002).

Romero de Pablos, Ana, and María Jesús Santesmases (coords), *Cien años de política científica en España* (Bilbao: Fundación BBVA, 2008).

Rossiter, Margaret W., *Women Scientist in America: Before Affirmative Action, 1940–1972* (Baltimore MD: Johns Hopkins University Press, 1995).

Ruiz Bautista, E., 'En pos del "buen lector": censura editorial y clases populares durante el primer franquismo', *Espacio, Tiempo y Forma, Serie V, Historia Contemporánea,* 16 (2004), 231–51.

— *Los señores del libro: propagandistas, censores y bibliotecarios en el primer franquismo (1939–1945)* (Gijón: Trea, 2005).

Sales, Joaquim, *Enrique Moles: una biografía científica y política* (Barcelona and Madrid: Universitat de Barcelona-CSIC, 2021).

Sánchez Ron, José Manuel, *Cincel, martillo y piedra. Historia de la ciencia en España (siglos XIX y XX)* (Madrid: Taurus, 1999).

— *El Consejo Superior de Investigaciones Científicas. Una ventana al conocimiento (1939–2014)* (Madrid: CSIC, 2021).

Santesmases Navarro de Palencia, María Jesús, *Establecimiento de la bioquímica y de la biología molecular en España (1940–1970)* (Madrid: CSIC, 1997).
— *Mujeres científicas en España (1940–1970). Profesionalización y modernización social* (Madrid: Instituto de la Mujer, 2000).
Santesmases Navarro de Palencia, María Jesús, and Emilio Muñoz Ruiz, *Los primeros años del Consejo Superior de Investigaciones Científicas. Una introducción a la política científica del régimen franquista* (Madrid: Documento de Trabajo, 1993).
Santoveña Setién, Antonio, *Menéndez Pelayo y las derechas en España* (Santander: Librería Estudio, 1994).
Sanz Menéndez, Luis, *Estado, ciencia y tecnología en España: 1939–1997* (Madrid: Alianza, 1997).
Shapin, S., 'Discipline and Bounding: The History and Sociology of Science as Seen Through the Externalism-Internalism Debate', *History of Science*, 30 (1992), 333–69.
— *The Scientific Life: A Moral History of a Late Modern Vocation* (Chicago IL: University of Chicago Press, 2008).
— *Never Pure: Historical Studies of Science as if It Was Produced by People with Bodies, Situated in Time, Space, Culture, and Society, and Struggling for Credibility and Authority* (Baltimore MD: Johns Hopkins University Press, 2010).
Valera Candel, Manuel, and Carlos López Fernández, *La física en España a través de los Anales de la Real Sociedad Española de Física y Química, 1903–1965* (Murcia: Universidad de Murcia, 2001).
Viñas, Ángel, *En las garras del águila. Los pactos con Estados Unidos, de Francisco Franco a Felipe González (1945–1995)* (Barcelona: Crítica, 2003).
— (ed.), *El combate por la historia. La república, la guerra civil y el franquismo* (Barcelona: Pasado y Presente, 2012).

Notes

Introduction

1 Note of 31 August 1943; Rocasolano Library Archive, Carpeta Libros 1943.
2 *CSIC, Estructura y norma de la investigación Nacional* (Madrid: CSIC, 1940), 65–8.
3 Note of 31 August 1943; Rocasolano Library Archive, Carpeta Libros 1943.
4 *CSIC, Memoria de la Secretaría General (1943)* (Madrid: CSIC, 1944), 313. For the subsequent citations from the official Spanish National Research Council annual activity reports: *CSIC*, year of publication, and page number for the reference.
5 Note of 31 August 1943; Rocasolano Library Archive, Carpeta Libros 1943.
6 List of packages sent from the Alonso Barba Institute, National Institute of Chemistry, 30 [?] August 1943; Rocasolano Library Archive, Carpeta Libros 1943.
7 Letter from José María Albareda to José Lorente (undersecretary, Ministry of Government), 29 April 1940; AGUN / JMA / 006 / 001 / 0011.
8 Letter from José María Albareda to José Lorente (undersecretary, Ministry of Government), 29 April 1940; AGUN / JMA / 006 / 001 / 0011.
9 *CSIC*, 1943, 52.
10 *CSIC*, 1942, 2.
11 Rocasolano Library Archive, Carpeta Índice de los trabajos realizados en los Institutos de Física y Química, 1931–46.
12 *CSIC*, 1942, VI (Prologue).
13 Order of 18 February 1941, *BOE*, 28 February 1941.
14 'El libro español en manos españolas. Las publicaciones del Consejo Superior de Investigaciones Científicas mostrarán, en el Día del Libro, el trabajo de los Patronatos e Institutos nacionales', *ABC*, 20 April 1941.

Chapter 1

1 *CSIC, Memoria de la Secretaría General (1949)* (Madrid: CSIC, 1951), 77. For the subsequent citations from the official Spanish National Research Council annual activity reports: *CSIC*, year of publication, and page number for the reference.
2 *CSIC*, 1942, 118.
3 *CSIC*, 1951, 78.
4 *CSIC*, 1942, 111.

Chapter 2

1 *CSIC, Memoria de la Secretaría General (1946–1947)* (Madrid: CSIC, 1948), 15. For the subsequent citations from the official Spanish National Research Council annual activity reports: *CSIC*, year of publication, and page number for the reference.
2 *CSIC*, 1942, 29.
3 José Ibáñez Martín, *El sentido político de la Cultura en la hora presente* (Madrid, [n.p.], 1942), p. 10.
4 *CSIC*, 1959, 174.

Chapter 3

1 *CSIC, Memoria de la Secretaría General (1942)* (Madrid: CSIC, 1943), 51–2. For the subsequent citations from the official Spanish National Research Council annual activity reports: *CSIC*, year of publication, and page number for the reference.
2 *CSIC*, 1942, 29.
3 Manuel Lora-Tamayo, 'Recuerdos del Consejo Superior de Investigaciones Científicas en su 50° Aniversario', *Arbor: ciencia, pensamiento y cultura*, 529, January (1990), 107.
4 *CSIC, Estructura y norma de la investigación Nacional* (Madrid: CSIC, 1940), 11–12.
5 Manuel Pombo Angulo, 'Monseñor Albareda', *La Vanguardia*, 29 March 1966.
6 'Un testimonio sobre el Consejo Superior de Investigaciones Científicas', *Revista Nacional de Educación*, 64 (1946), 93–4.
7 *CSIC*, 1940, 9–10.
8 *CSIC*, 1942, 11–28.
9 *CSIC*, 1950, 37.
10 *CSIC*, 1942, 28–52.
11 *CSIC*, 1942, 30–1.
12 *CSIC*, 1945, 42–3.
13 José Ibáñez Martín, 'Un año de política docente', *Revista Nacional de Educación*, 10 (1941), 8–13.

14 *CSIC*, 1942, 33.
15 *CSIC*, 1948, 135.
16 *CSIC*, 1943, 45–6.
17 *CSIC*, 1943, 50.
18 *CSIC*, 1948, 142.
19 *CSIC*, 1942, 32.
20 *CSIC*, 1948, 54.
21 *CSIC*, 1942, 32.
22 *CSIC*, 1944, 10.
23 *CSIC*, 1944, 9–10.
24 *CSIC*, 1948, 142.
25 *CSIC*, 1942, 47.
26 *CSIC*, 1942, 34.
27 *CSIC*, 1942, 114–15.
28 *CSIC*, 1943, 1.
29 *CSIC*, 1942, 6.
30 *CSIC*, 1942, 48.

Chapter 4

1 *CSIC, Memoria de la Secretaría General (1943)* (Madrid: CSIC, 1944), 49–50. For the subsequent citations from the official Spanish National Research Council annual activity reports: *CSIC*, year of publication, and page number for the reference.
2 *CSIC*, 1944, 58.
3 *CSIC*, 1950, 50.
4 *CSIC*, 1950, 50.
5 *CSIC*, 1942, 115.
6 Manuel García Doncel, 'Filosofía y Ciencias. Filósofos y Científicos a la luz de las alocuciones de Pío XII', *Espíritu*, 7 (1956), 202–12.
7 Jaime Echarri Goicoechea, 'Pío XII y la Filosofía Perenne ante las ciencias', *Salamanticensis*, 3 (1956), 321–49.
8 *CSIC*, 1948, 156.
9 *CSIC*, 1944, 57.
10 *CSIC*, 1950, 36.
11 *CSIC*, 1948, 142.
12 *CSIC*, 1948, 142.
13 Echarri Goicoechea, 'Pío XII y la Filosofía Perenne ante las ciencias', 326.
14 *CSIC*, 1959, 63.
15 *CSIC*, 1959, 64
16 *CSIC*, 1959, 173–4.
17 *CSIC*, 1959, 173.

18 *CSIC, Hacia una nueva organización científica en los Estados Unidos* (Madrid: CSIC-Patronato Juan de la Cierva Codorníu, 1947), 7.
19 *Special Message to the Congress Presenting a 21-Point Program for the Reconversion Period*, 6 September 1945.
20 *CSIC*, 1946, 75–6.
21 *CSIC*, 1942, 5.
22 *CSIC*, 1951, 66–8.
23 *CSIC*, 1951, 88.
24 Inaugural speech at the CSIC Biological Research Centre, 8 February 1958.
25 *CSIC*, 1952, 93.
26 *CSIC*, 1952, 92.
27 *CSIC*, 1959, 167–9.
28 *CSIC*, 1952, 78.
29 *CSIC*, 1952, 95–6.
30 *CSIC*, 1942, 7.
31 *CSIC*, 1952, 34.
32 *CSIC*, 1952, 88–91.
33 *CSIC*, 1959, 58.
34 *CSIC*, 1959, 68.
35 Letter from José Royo to José Ibáñez Martín, 14 May 1959; AGUN/JIM/590/R-7.
36 Letter from José Royo to José Ibáñez Martín, 14 May 1959; AGUN/JIM/590/R-7.
37 Letter from José Ibáñez Martín to José Royo, 23 May 1959; AGUN/JIM/590/R-7.
38 Letter from Alberto Sols to José Ibáñez Martín, 4 February 1960; AGUN/JIM/590/S-21.
39 Letter from Alberto Sols to José Ibáñez Martín, 4 February 1960; AGUN/JIM/590/S-21.

Chapter 5

1 *CSIC, Memoria de la Secretaría General (1949)* (Madrid: CSIC, 1951), 91. For the subsequent citations from the official Spanish National Research Council annual activity reports: *CSIC*, year of publication, and page number for the reference.
2 *CSIC*, 1951, 92.
3 'Wherefore, I affirm to you, Spanish intellectuals . . .. "Intellectuals, who hear me!" The Fatherland, which wishes to be great again, calls you with an anguished cry'; *CSIC*, 1942, 38 and 41.
4 *CSIC*, 1942, 38–41.
5 *CSIC*, 1942, 39.

6 *CSIC*, 1942, 110.
7 *CSIC*, 1942, 110–11.
8 *CSIC*, 1942, 111.
9 Cited in José María Albareda Herrera, *Consideraciones sobre la investigación científica* (Madrid: CSIC, 1951), p. 21.
10 José Ibáñez Martín, 'Albareda al servicio de España'; AGUN/JIM/596/3. It is little wonder that Albareda was ordained as an Opus Dei priest.
11 *CSIC*, 1946, 66.
12 Letter from José Ibáñez Martín to CSIC Vice President José Royo, 23 May 1959; AGUN/JIM/590/R-7.

Chapter 6

1 *CSIC, Colaboradores e investigadores del Consejo Superior de Investigaciones Científicas* (Madrid: CSIC, 1956), 7.
2 CSIC, *Colaboradores e investigadores*, 3.
3 CSIC, *Colaboradores e investigadores*, 9–14.
4 CSIC, *Colaboradores e investigadores*, 8.
5 CSIC, *Colaboradores e investigadores*, p. 8.
6 Letter from María Dolores Angulo to the head of CSIC publications, 8 June 1954; CSIC Editorial Archive, caja 29, expdte. 1043.
7 CSIC, *Colaboradores e investigadores*, 100.
8 CSIC, *Colaboradores e investigadores*, 278.

Chapter 7

1 *BOE*, 11 November 1940.
2 *BOE*, 15 March 1946.
3 *BOE*, 20 June 1946.
4 *BOE*, 15 March 1946.
5 An application to work on nuclear atomic structures and problems related to disintegration and atomic energy; *BOE*, 20 June 1946.
6 An application for a four-month fellowship to work in England on modern psychiatric treatment for substance abuse; *BOE*, 20 June 1946.
7 *BOE*, 20 June 1946.
8 Women represented 3.8 per cent of all the board's fellows. After 1957, they were joined by María Teresa Santamaría Redondo, who received her fellowship in 1956 to study at the University of Wisconsin-Madison in the United States for an unidentified period; Olga Moreiras Tuni, who received a fellowship for three months in 1960 to work in the Bromatology Laboratory at the Universiteit Utrecht; María del Carmen Tornero García, who was granted a nine-month fellowship

in 1960 to work at the Universität zu Köln in Germany; and Gertrudis Torrontegui y Pico de Coaña, granted a one-year fellowship in July 1959 at the Universiteit Utrecht; see *CSIC, Becarios del Consejo Superior de Investigaciones Científicas 1957–1960* (Madrid: CSIC, 1961), 23–34.

9 Women received 16 per cent of all the fellowships granted by this board. In addition to the women described in the text, the following female scientists also received fellowships to study abroad: Trinidad Aleixandre Campos, 1959, Universidad de Lisboa, Universidad de Coimbra and Universidad do Porto; María Pilar Arévalo Carretero, 1957, nine months with Professor F. Smithson, University College of North Wales-Bangor; Josefina Benayas Casares, 1957, seven months with Professor Smithson in England; Eulalia Cabezas de Herrera Sánchez, 1960, one month, Institut Pasteur; Filomena Díaz Celayeta, 1957, one month with Professor Johanna in Westerdijk, Netherlands and 1959, three months with Antonio Branquinho in Portugal; Ana María García Gómez, 1960, six months with Professor Cámara in Portugal; María del Carmen Gil Fernández, 1956, three months, University of Cambridge, England; María Luz Ibáñez Martín, 1959, ten months with Professor Mirimanoff, Université de Genève, Switzerland; Marina Kress Voltz, 1960, two months with Professor Kubiena, Germany; Asunción Linares Rodríguez, 1959, three months, France; Matilde Martínez Vázquez, 1959, with Professor Sousa de Cámara and Professor Azevedo Coutinho, Portugal; Montserrat Miquel Berenguer, 1957, four months with Professor Sommerville, England; Josefina Ogando Rubio, 1960, three months with Professor Strenzke, Germany; Ángeles Peiró Callizo, 1960, two months, France; Pilar Sánchez Conde, 1960, three months with Professor Mückenhausen, Germany; Irene Sánchez Vega Castillo, 1959, two months with Professor Guido Carobbi, Italy; Dolores Selga Serra, 1960, four months with Professor H. Franz and Professor W. Kühnelt, Switzerland and Germany; Baldomera Velázquez Sánchez, 1956, eight months in France; and Carmen Virgili, 1960, a conference in Denmark; see CSIC, *Becarios 1957–1960*, 35–60.

10 Alfonso V. Carrascosa Santiago, 'Católicas y científicas: Josefa Menéndez Amor', *Revista Ecclesia*, 12 February 2018, *www.revistaecclesia.com/catolicas-y-cientificas-josefa-menendez-amor-por-alfonso-v-carrascosa-cientifico-del-csic/* (last accessed 13 April 2022).

11 *CSIC, Memoria de la Secretaría General (1946–1947)* (Madrid: CSIC, 1948), 499. For the subsequent citations from the official Spanish National Research Council annual activity reports: *CSIC*, year of publication, and page number for the reference.

12 *BOE*, 3 July 1971.

13 *BOE*, 3 July 1971.

14 *M.ª Concepción Bonet, geóloga honorífica a título póstumo*: *https://cgeologos.es/noticia/m-concepcion-bonet-geologa-honorifica-a-titulo-postumo* (last accessed 13 April 2022).

15 Women represented 6.2 per cent of all the board's fellows. In addition to the women described in the text, the following female scientists also received fellowships to study abroad: Pilar Alonso Sanjuan, 1960, one year with Professor Raymond Pepinsky, University of Pennsylvania, United States; María Emilia García Clavel, 1960, a grant to attend the Fourth International Symposium on the Reactivity of Solids, Amsterdam, Netherlands, 30 May–4 June; Amparo López Castro, 1959 with Professor Cox, School of Chemistry, University of Leeds, England; María del Carmen Martín Casasempere, 1960, three months with Professor Stackelberg, Universität Bonn, Germany; Isabel Martín Tordesillas, 1960, one year with Professor WFK Wynnes-Jones, England; Sagrario Martínez Carrera, 1959, one year with Dr McGillavry, Netherlands; Trinidad Marzo Muñoz-Cobo, 1960, one-and-a-half months with Professor Lederer, France; Josefa María Molera Mayo, 1959, fifteen days to visit the laboratories of the Department of Chemistry, University of Sheffield, England; Felisa Nuñez Cubero, 1959, one year with Professor L. F. Bates, University of Nottingham, England; Mercedes Pérez Fernández, 1959, ten months with Professor Gerischer, Germany; and Carmen Pino Vázquez, 1959, four months with Professor Goubeau, Stuttgart, Germany; see CSIC, *Becarios 1957–1960*, 61–74.

16 *BOE*, 28 July 1945.

17 Bryn Mawr College, *Report of the President to the Board of Directors (1949–1950)*, December 1950, p. 27.

18 *BOE*, 3 July 1971.

19 These women represented 6.8 per cent of the board's fellows. Between 1957 and 1960, they were joined by Laura Iglesias Romero, who received outside funding in July 1958 to study for a year at Princeton University; see *CSIC, Becarios 1957–1960*, 75–8.

20 Transcript of an interview with Piedad de la Cierva and Carmen González Farinós conducted by Néstor Herran; CEHIC, UAB (Madrid, 8 November 2004), 3–4.

21 *Juana Bellanato Fontecha o la voluntad de ser científica*: *http://161.111.98.22/web/guest/juana-bellanato* (last accessed 10 June 2020).

22 *BOE*, 15 April 1961.

23 *CSIC*, 1942, 48.

24 This number rises to eighty-four when the women who received fellowships between 1957 and 1960 are included.

25 These numbers increase when other types of fellowship are taken into consideration, such as funding from American universities;

see Paloma Alcalá Cortijo and Carmen Magallón Portolés, 'Avances, rupturas y retrocesos: mujeres en las ciencias experimentales en España (1907–2005)', in Ana Romero de Pablos and María Jesús Santesmases (coords), *Cien años de política científica en España* (Bilbao: Fundación BBVA, 2008), p. 149.

26 *CSIC*, 1943, 47.

Chapter 8

1 *Revista de Ciencia Aplicada*, 27 / VI, fasc. 4 (July–August 1952), 331–46.

2 'La primera mujer catedrática de una Escuela Técnica Superior', *ABC*, 28 April 1977, p. 22; and *BOE*, 2 July 1977.

3 These were published previously without González in *Revista de Ciencia Aplicada*, 19 and 21 / V, fascs 2 and 4, 143–9 and 330–5.

4 *Revista de Ciencia Aplicada*, 40 / VIII, fasc. 5 (September–October 1954), 385–92.

5 Footnote in *Revista de Ciencia Aplicada*, 48 / X, fasc. 1 (January–February 1956), 30–3.

6 *Felisa Núñez Cubero. Una pionera en física experimental: https://mujeresconciencia.com/2018/04/04/felisa-nunez-cubero-una-pionera-en-fisica-experimental/* (last accessed 13 April 2022).

7 *Revista de Ciencia Aplicada*, 73 / XIV, fasc. 2 (March–April 1960), 97–106.

8 'Las mujeres en la Universidad Politécnica de Madrid', exhibition '(1910–2010) Mujer y Universidad, un siglo de vida "La Mujer en la Universidad, la Ingeniería, la Arquitectura y el Deporte"', November 2011.

9 *Revista de Ciencia Aplicada*, 48 / X, fasc. 1 (January–February 1956), 13–21.

10 *CSIC, Memoria de la Secretaría General (1960)* (Madrid: CSIC, 1963), 462.

11 *Fermín Garrido Quintana: http://dbe.rah.es/biografias/38519/fermin-garrido-quintana* (last accessed 13 April 2022).

12 *CSIC, Becarios del Consejo Superior de Investigaciones Científicas* (Madrid: CSIC, 1957), 108.

13 *Revista de Ciencia Aplicada*, 49 / X, fasc. 2 (March–April 1956), 136–43. José María Garrido Márquez also appears after Concepción Llaguno Marchena in two other articles on a similar topic: 'Aspectos bioquímicos de la producción de ácido fumárico por fermentación', *Revista de Ciencia Aplicada*, 76 / XIV, fasc. 5 (September–October 1960), 385–96; and 'Producción de Ácido fumárico por fermentación. Estudio de la aireación y agitación del medio de cultivo, *Revista de Ciencia Aplicada*, 83 / XV, fasc. 6 (November–December 1961), 481–8.

14 'Concepción Llaguno, científica y gran comunicadora', *El País*, 9 October 2010: *http://elpais.com/diario/2010/10/09/necrologicas/1286575201_850215.html* (last accessed 13 April 2022).

15 Later in her career, she worked as a research professor at the CSIC after 1971, where she also held the position of vice-secretary general for scientific affairs, edited the oenology books *Enología: Temas actuales* (Asociación Nacional de Químicos, 1982) and *Guida di Vini d'Espagna* (Mondadori, 1982) and co-authored, with Carmen Polo, *El Vinagre de Vino* (CSIC, 1991), which received an award from the International Vine and Wine Office in 1993. She also received a prize for food science research in 1991, collaborated on the draft of the first National R&D Plan and headed the national food technology programme. In 1994, she received the 'Progressive Woman' award; *Concepción Llaguno Marchena (1925–2010): www.csic.es/es/el-csic/mujeres-y-ciencia/mujeres-ilustres/concepcion-llaguno-marchena* (last accessed 13 April 2022).

16 Margarita Buendía and José M.ª Garrido, 'Estudios sobre la producción del ácido itacónico', *Revista de Ciencia Aplicada*, 65/XII, fasc. 6 (November–December 1958), 481–95.

17 Margarita Buendía and José M.ª Garrido, 'La producción industrial del ácido itacónico', *Revista de Ciencia Aplicada*, 66/XIII, fasc. 1 (January–February 1959), 23–33; and 'Producción de ácido itacónico en cultivo sumergido usando un fermentador', *Revista de Ciencia Aplicada*, 67/XIII, fasc. 2 (March–April 1959), 130–40.

18 *Revista de Ciencia Aplicada*, 96/XVIII, fasc. 1 (January–February 1964), 11–22; 97/XVIII, fasc. 2 (March–April 1964), 140–53; 98/XVIII, fasc. 3, (May–June 1964), 234–46; 99/XVIII, fasc. 4, (July–August 1964), 330–41; 100/XVIII, fasc. 5 (September–October 1964), 430–9; 101/XVIII, fasc. 6 (November–December 1964), 498–513; 103/XIX, fasc. 2 (March–April 1965), 97–104; 104/XIX, fasc. 3 (May–June 1965), 217–25 and 105/XIX, fasc. 4 (July–August 1965), 316–23.

19 María Dolores Cabezudo read her doctoral thesis 'Aspectos de la crianza de vinos finos por levaduras de flor' in 1966.

20 M. J. Fernández, María Dolores Cabezudo, Concepción Llaguno and José María Garrido, 'Determinación del metanol en alcoholes vínicos e industriales, aguardientes y licores españoles', *Revista de Ciencia Aplicada*, 96/XVIII, fasc. 1 (January–February 1964), 30–7. In another co-authored article, Baldomero Íñigo Leal, the head of the Microbiology Laboratory, appeared after María Josefa Fernández and Concepción Llaguno; see María Josefa Fernández, Concepción Llaguno and Baldomero Íñigo, 'Metabolismo de Levaduras', *Revista de Ciencia Aplicada*, 98/XVIII, fasc. 3 (May–June 1964), 223–33.

21 In the 1950s, Fernández Alonso studied abroad under Raymond Dauder, secretary of the Centre de Mécanique Ondulatoire de

France in Paris, and at the California Institute of Technology, under the direction of Prof Linus Pauling, shortly before he received the Nobel Prize in chemistry in 1954: *www.uv.es/uvweb/rectorado/es/premios-distinciones/doctores-honoris-causa/doctores-honoris-causa-siglo-xx/fecha-investidura/acto-investidura-doctor-honoris-causa-del-excm-sr-dr-jose-ignacio-fernandez-alonso-1285872205498/Honoris.html?id=1285878595903* (last accessed 13 April 2022).

22 *Revista de Ciencia Aplicada*, 60/XII, fasc. 1 (January–February 1958), 26–30.

23 *Revista de Ciencia Aplicada*, 54/XI, fasc. 1 (January–February 1957), 25–31. She was also the sole author of 'Estudio de la oxidación peryódica de los ácidos málico, cítrico y tartiárico', *Revista de Ciencia Aplicada*, 100/XVIII, fasc. 5 (September–October 1964), 412–15.

24 Carlos Gómez Herrera and Carmen Janer del Valle, 'Problemas físico-químicos en la extracción del aceite de oliva', *Revista de Ciencia Aplicada*, 55/XI, fasc. 2 (March–April 1957), 97–103.

25 *Revista de Ciencia Aplicada*, 60/XII, fasc. 1 (January–February 1958), 16–25.

26 *Revista de Ciencia Aplicada*, 84/XVI, fasc. 1 (January–February 1962), 36–41.

27 The instrument was built by M. Guibert, a mechanical designer in the Department of Physical Crystallography; see *Revista de Ciencia Aplicada*, 88/XVI, fasc. 5 (September–October 1962), 385–96. In 1961, the same funding was behind the publication of an article by Amorós, Guibert, Canut and Arrese, 'Un goniómetro Weissenberg vertical para altas y bajas temperaturas', *Revista de Ciencia Aplicada*, 81/XV, fasc. 4 (July–August 1961), 289–97.

28 *Revista de Ciencia Aplicada*, 92/XVII, fasc. 3 (May–June 1963), 199–213. Carbonell and Canut – in that order – published another work in 1964: A Carbonell and M. L. Canut, 'Determinación del espectro elástico vibracional de un cristal molecular, la hexamina, por difracción difusa térmica de rayos X', *Revista de Ciencia Aplicada*, 97/XVIII, fasc. 2 (March–April 1964), 126–39.

29 *Revista de Ciencia Aplicada*, 45/IX, fasc. 4 (July–August 1955), 289. Years later, in 1966, Piedad de la Cierva Viudes and her collaborators, Guadalupe Ortiz de Landázuri and Antonia Muñoz, once again won the Juan de la Cierva Prize for their work *Estudio físico-químico y posibilidades industriales de las cenizas de la cascarilla de arroz*; see Dolores Sánchez Durá and Vicenta Verdugo Martí, *Acceso de las mujeres a la Universitat de Valencia (1910–1960)* (Valencia: Unitat d'Igualtat, 2011), p. 51.

30 'Although the real director was Piedad de la Cierva, she had to ask José María Albareda, who was formally the director, as Piedad was

not a chair professor'; see Inmaculada Alva Rodríguez, 'Piedad de la Cierva: una sorprendente trayectoria profesional durante la Segunda República y el Franquismo', *Arbor*, 192/779 (May–June 2016), 9.

31 *Revista de Ciencia Aplicada*, 63/XII, fasc. 4 (July–August 1958), 289–92. This was a continuation of an article published in English by the two women in the *American Ceramic Society Bulletin*, 35/191 (1956), titled 'Properties of Some Spanish Clay and their Suitability for Optical Glass Pots. I'.

32 *Revista de Ciencia Aplicada*, 68/XIII, fasc. 3 (May–June 1959), 227–33.

33 *Revista de Ciencia Aplicada*, 99/XVIII, fasc. 4 (November–December 1964). She had published once before in *Revista de Ciencia Aplicada*, 56/XI, fasc. 3 (May–June 1957).

34 Fernando Burriel Martí and Carmen Álvarez Herrero, 'Determinación complexométrica de hierro, aluminio, calcio y magnesio aplicable a los cementos de arcilla', *Revista de Ciencia Aplicada*, 106/XIX, fasc. 5 (September–October 1965), 396–401; and 111/XX, fasc. 4 (July–August 1966), 308–14.

35 *Revista de Ciencia Aplicada*, 112/XX, fasc. 5 (September–October 1966), 385–94.

36 As a CSIC collaborator, she also received the prestigious Ramsay Memorial Fellowship Trust in 1950 to work in the Physical Chemistry Laboratory at the University of Oxford under Professor Sir C. N. Hinshelwood, the 1956 Nobel Prize winner for chemistry. She built what must have been the first gas chromatograph in Spain and received, among other awards, the CSIC Alfonso X El Sabio Prize (1966) and the Perkin-Elmer Prize (1967), in addition to being awarded the Chemistry Medal from the Spanish Royal Society of Physics and Chemistry. Among other distinctions, she was president of the Spanish Society of Chromatography and Related Techniques, which she founded in 1973; see Santiago Carrascosa, 2012a, pp. 383–4.

37 Thus, during this twenty-year period, women comprised 10.8 per cent of the total number of authors and co-authors; see Fernando García Naharro, *Palabras, actores, objetos y lugares. El CSIC y la demarcación de la ciencia en la España franquista (1939–1966)* (Madrid: CSIC, 2022), pp. 303–34.

Chapter 9

1 Letter from Juan Roger to Rafael de Balbín, 26 January 1954; CSIC Editorial Archive, caja 39, expdte. 1656.

2 Letter from Juan Roger to Rafael de Balbín, 17 February 1954; CSIC Editorial Archive, caja 39, expdte. 1656.

3 CSIC Editorial Archive, caja 3, expdte. 1656.

4 *Consejo Superior de Investigaciones Científicas*, Spain, 1954, p. 27.
5 Presentation on publications; AGUN / JIM / 592 / 5.
6 *CSIC, Memoria de la Secretaría General (1949)* (Madrid: CSIC, 1951), 87. For the subsequent citations from the official Spanish National Research Council annual activity reports: *CSIC*, year of publication, and page number for the reference.
7 *CSIC*, 1945, 57.
8 Letter from Rafael de Balbín to José Ibáñez Martín, 26 January 1960; AGUN / JIM / 587 / B-4 (1960).
9 AGUN / JIM / 589 / M-3 (1960–8).
10 AGUN / JIM / 587 / C-21 (1961).
11 AGUN / JIM / 587 / C-21 (1961).
12 Lisbon, 23 October 1961, AGUN / JIM / 587 / C-21 (1961).
13 Letter of 13 November 1961, AGUN / JIM / 588 / G-7 (1961).
14 *Boletín Bibliográfico del Consejo Superior de Investigaciones Científicas*, 1 (1944), 1.
15 Data from 1944 to 1950 taken from the *Boletín Bibliográfico del Consejo Superior de Investigaciones Científicas*, 1–63, 1944–51.
16 Subject index, 1954 General Catalogue; CSIC Editorial Archive, caja 45, expdte. 1754.
17 'The designation of Royal was granted to the Society by Alfonso XIII on the occasion of the celebration of the Silver Weddings, but was removed during the Republican era'; see Manuel Valera Candel and Carlos López Fernández, 2001, p. 195, note 451.
18 Words from the minutes of the meeting of 15 November 1939, in Manuel Valera Candel and Carlos López Fernández, 2001, pp. 199–201.
19 *CSIC*, 1948, 139–40.
20 Meeting of 25 November 1962; see Manuel Valera Candel and Carlos López Fernández, 2001, pp. 230 and 232, note 514.
21 Letters from 19 September 1941; Rocasolano Library Archive, Carpeta Libros 1941.
22 *Boletín Bibliográfico del Consejo Superior de Investigaciones Científicas*, 6 (1944), 10.

Chapter 10

1 *BOE*, 26 February 1949.
2 Circular of 8 June 1940; Archivo Biblioteca Rocasolano, Carpeta Libros, 1940.
3 *Reglamento de las publicaciones del Consejo Superior de Investigaciones Científicas*, approved by the Permanent Commission of the Raimundo Lulio, Menéndez Pelayo and Diego de Saavedra Fajardo Boards, 24 June 1953; CSIC Editorial Archive, caja 45, expdte. 1746.

4 Note proposing the creation of a board to monitor CSIC investments; AGUN / JIM / 592 / Complejidad económica (n.d.).
5 Letter from Antonio Rius Miró to José Ibáñez Martín, 9 November 1953; AGUN / JIM / 592 / 3.
6 Letter of 9 November 1953; AGUN / JIM / 592 / 3.
7 Letter from Gregorio Marañón to José María Albareda, 20 July 1956; AGUN / JIM / 586 / A-3 (1956).
8 'Regulations for the distribution of free issues', 13 March 1957, AGA. Educación. Fondo CSIC, caja 31 / 8882, carpeta Sección de Publicaciones, 1957.
9 *CSIC, Memoria de las actividades desarrolladas por el Patronato 'Juan de la Cierva' de Investigación Técnica. Año 1950* (Madrid: CSIC, 1951), 18.
10 *CSIC, Memoria de las actividades desarrolladas por el Patronato 'Juan de la Cierva' de Investigación Técnica*, 20.

Chapter 11

1 The subtitle of this chapter is a paraphrase of an expression by Robert Darnton. A true master of the history of books, his writing has greatly influenced my scholarship. Robert Darnton, 'What is the History of Books?' revisited, *Modern Intellectual History*, 4 / 3 (2007), 495–508.
2 See *Revista de Ciencia Aplicada*, 55 / XI / fasc. 2 (March–April 1957), 104.
3 *CSIC, Memoria de las actividades desarrolladas por el Patronato 'Juan de la Cierva' de Investigación Técnica. Año 1946* (Madrid: CSIC, 1947), 26. For the subsequent citations from the official Juan de la Cierva Board annual activity reports: *CSIC*, year of publication, and page number for the reference.
4 *Revista de Ciencia Aplicada*, 1 / I / fasc. 1 (October–December 1947), 56.
5 *CSIC*, 1948, 28.
6 *Revista de Ciencia Aplicada*, 1 / I / fasc. 1 (October–December 1947), 51.
7 *CSIC*, 1949, 8.
8 *Revista de Ciencia Aplicada*, 6 / III / fasc. 1 (January–February 1949), 69.
9 *CSIC*, 1960, 10.
10 *CSIC*, 1959, 8.
11 *CSIC*, 1955, 41.
12 *Revista de Ciencia Aplicada*, 41 / VIII / fasc. 6 (November–December 1954), 542.
13 *CSIC*, 1951, 29.
14 *CSIC*, 1965, 14.
15 *Revista de Ciencia Aplicada*, 12 / IV / fasc. 1 (January–February 1950), 84, footnote.
16 *Revista de Ciencia Aplicada*, 12 / IV / fasc. 1 (January–February 1950), 84, footnote.

17 *CSIC*, 1962, 29.
18 *CSIC*, 1965, 79.
19 Letter from the director of the Institute for Cement to Rafael de Balbín, CSIC undersecretary, 26 October 1947, AHT/OI/A/003/119.
20 Meeting of 13 April 1948, AHT/OI/A/003/200.
21 *CSIC*, 1950, 39.
22 *CSIC*, 1965, 83.
23 *CSIC*, 1954, 41.
24 *CSIC*, 1950, 51.
25 *CSIC*, 1955, 63.
26 *CSIC*, 1950, 51.
27 However, other journals like *Grasas y Aceites* also translated their articles into different languages.
28 *CSIC*, 1949, 66.
29 *CSIC*, 1949, 66.
30 *CSIC*, 1949, 68.
31 *CSIC*, 1949, 66.
32 *CSIC*, 1953, 36.
33 *CSIC*, 1954, 40.
34 *CSIC*, 1955, 69.
35 *CSIC*, 1956, 43.
36 Analysis from the Official Research Centre, Madrid, March 1972, 60; AGA (5)4 31/12.327.
37 *CSIC*, 1954, 38.
38 *CSIC*, 1956, 37; 41.
39 *CSIC*, 1958, 33.
40 *CSIC*, 1962, 35.
41 *CSIC*, 1953, 52.
42 *CSIC*, 1966, 75–6.
43 *CSIC*, 1965, 65.
44 *CSIC*, 1948, 51.
45 *CSIC*, 1952, 72.
46 The main characters in these drawings were always men who were either interested in the institute's information or in the scientific work being done there, which was represented in the sketches of male subjects in laboratories fitted out with test tubes, pipes and various other gadgets; *Boletín Informativo del Instituto Nacional del Carbón*, 1 (1952) and 20 (1955).
47 *CSIC*, 1953, 40.
48 *Boletín informativo del Instituto Nacional del Carbón*, Instituto Nacional del Combustible, Oviedo, 11 (September 1953), 39.
49 *CSIC*, 1956, 45.
50 *CSIC*, 1954, 44.

51 *CSIC*, 1955, 72.
52 *CSIC*, 1957, 52.
53 *CSIC*, 1960, 44.
54 *CSIC*, 1955, 72.
55 According to the sources, in 1952, ten translations and 150 photocopy reproductions were provided; in 1953, thirty-six translations and 'approximately' 200 photocopy reproductions; and in 1954, 250 'requests' for reproductions; CSIC, 1952, 1953 and 1954, 40, 44 and 72, respectively.
56 These are 1954 prices; see *Boletín informativo del Instituto Nacional del Carbón*, Instituto Nacional del Combustible, Oviedo, 18 (November 1954), 1.
57 Vicente Gómez Aranda, 'Luis Bermejo Vida', *Combustibles*, CSIC-Instituto del Combustible, 1/5 (1941), 180.
58 Single issues sold for 6 pesetas in Spain, 7 pesetas in Portugal and Latin America and 8 pesetas in other foreign countries; *Combustibles*, CSIC-Instituto del Combustible, 1/1 (1941), 52.
59 In 1946, when the Zaragoza Institute of Fuel became a national institute, the laboratory in that city, located in the basement of the Faculty of Medicine, was designated a section; 'Crónica del Instituto del Combustible', *Combustibles*, CSIC-Instituto del Combustible, 1/1 (1941), 52; *CSIC*, 1965, 71.
60 *Combustibles*, CSIC-Instituto del Combustible, 1/3 (1941), 138.
61 *Combustibles*, CSIC-Instituto del Combustible, 1/4 (1941), 179.
62 *Compañía Arrendataria del Monopolio de Petróleos, S. A.*, CAMPSA. *Combustibles*, CSIC-Instituto del Combustible, 1/6 (1941), 254.
63 *CSIC*, 1947, 54.
64 *CSIC*, 1948, 55.
65 *Combustibles*, CSIC-Instituto del Combustible, 9/50 (1949), 170.
66 An advertising brochure; *Combustibles*, CSIC-Instituto del Combustible, 10/53 (July–September 1950).
67 *Combustibles*, CSIC-Instituto del Combustible, 10/51. The back cover bookplate was also stylised in 1952.
68 Drawings of men of science, alone in the laboratory or consulting journals; once again apparatuses and journals provide the only company for the *vir modestus*; *Combustibles*, CSIC-Instituto del Combustible, 13/65 (January–February 1953).
69 *CSIC*, 1953, 52.
70 *Combustibles*, CSIC-Instituto del Combustible, 12/59 (January-February 1952), 3–12.
71 *CSIC*, 1954, 55.
72 *CSIC*, 1955, 84.

73 'The total price of the first and second part is 185 pesetas' (loose publicity); *Ciencia y Técnica de la Soldadura,* CSIC-Instituto de la Soldadura, 6/28 (January–February 1956).
74 *CSIC,* 1948, 106–8.
75 Editorial, 'Programa de Trabajos', *Ciencia y Técnica de la Soldadura,* CSIC-Instituto de la Soldadura, 8/45 (November–December 1958).
76 As seen in the first issue, which contains the library catalogue; *CSIC,* 1949, 112.
77 *CSIC,* 1950, 116.
78 *CSIC,* 1951, 154.
79 *CSIC,* 1952, 125.
80 The institute's aims and work were presented in the editorial that introduced the first issue of *Ciencia y Técnica de la Soldadura,* CSIC-Instituto de la Soldadura, 1/2 (September–October 1951).
81 Each collection contained ten filing folders (loose publicity); *Ciencia y Técnica de la Soldadura,* CSIC-Instituto de la Soldadura, 8/28 (January–February 1956).
82 *Ciencia y Técnica de la Soldadura,* CSIC-Instituto de la Soldadura, 1/2 (September–October 1951), 2 (bibliography) and 16 (science and technology).
83 Beginning with issue 6 (May–June 1952), the only drawings that appeared were explanatory, as well as the figures and charts that accompanied the research articles. When they reappeared in 1954, the figure of a man had been replaced by a young welder.
84 *CSIC,* 1955, 121.
85 *CSIC,* 1954, 82.
86 *Ciencia y Técnica de la Soldadura,* CSIC-Instituto de la Soldadura, 6/30 (May–June 1956).
87 Editorial, 'El número 50', *Ciencia y Técnica de la Soldadura,* CSIC-Instituto de la Soldadura 9/50 (September–October 1959).
88 *Ciencia y Técnica de la Soldadura,* CSIC-Instituto de la Soldadura, 3/15 (November–December 1953), last page.
89 Editorial, 'La propaganda al servicio de la ciencia', *Ciencia y Técnica de la Soldadura,* CSIC-Instituto de la Soldadura, 41 (March–April 1958).
90 *CSIC,* 1954, 82.
91 Editorial, 'Ideas técnicas y científicas', *Ciencia y Técnica de la Soldadura,* CSIC-Instituto de la Soldadura, 8/43 (July–August 1958).
92 Editorial, 'Ideas técnicas y científicas', *Ciencia y Técnica de la Soldadura,* CSIC-Instituto de la Soldadura, 8/43 (July–August 1958).
93 Chapter 4, Order of 14 March 1946. 'Creación', *Revista del Instituto del Hierro y del Acero,* CSIC-Instituto del Hierro y del Acero, 1/1 (July–September 1948), 6.

94 *CSIC*, 1948, 114.
95 'A nuestros miembros y asociados', *Revista del Instituto del Hierro y del Acero*, CSIC-Instituto del Hierro y del Acero, 1/1 (July–September 1948), 3–4.
96 'A nuestros miembros y asociados', *Revista del Instituto del Hierro y del Acero*, CSIC-Instituto del Hierro y del Acero, 1/1 (July–September 1948), 3–4.
97 'Organización', *Revista del Instituto del Hierro y del Acero*, CSIC-Instituto del Hierro y del Acero, 1/1 (July–September 1948), 9.
98 'Actuación', *Revista del Instituto del Hierro y del Acero*, CSIC-Instituto del Hierro y del Acero, 1/1 (July–September 1948), 10.
99 *Publicaciones*, Instituto del Hierro y del Acero, Madrid (January 1952), 19.
100 'Organización', *Revista del Instituto del Hierro y del Acero*, CSIC-Instituto del Hierro y del Acero, 1/1 (July–September 1948), 9.
101 'These works can be sent to the Institute in duplicate, and those that are published will be paid for on the basis of our official fees'; 'Nota Importante' *Revista del Instituto del Hierro y del Acero*, CSIC-Instituto del Hierro y del Acero, 2/3 (January–March 1949).
102 *Publicaciones*, Instituto del Hierro y del Acero, Madrid (January 1952), 18.
103 *Revista del Instituto del Hierro y del Acero*, CSIC-Instituto del Hierro y del Acero, ½ (October–December 1948), 9; and 2/3 (January–March 1949), 35.
104 Writing in May 1963 in an article published in the daily *YA*, José María Otero Navascués observed that, 'The nineteenth-century idea of the scholar, isolated in a corner of the laboratory or in his own home, able to make fundamental discoveries with rudimentary tools, is still quite widespread in significant circles'; see Carlos Pérez Fernández-Turégano, *José María Otero Navascués. Ciencia y Armada en la España del siglo* XX (Madrid: CSIC, 2012), p. 66. In this respect, in 1980, Pedro González Blasco asserted that 'in the mentality of Spanish society, the image of the scientist is still the old image of the "solitary man", and this has not been replaced by the new image of the scientist working in teams'; see Pedro González Blasco, *El investigador científico en España* (Madrid: CSIC, 1980), p. 47.
105 *Revista del Instituto del Hierro y del Acero*, CSIC-Instituto del Hierro y del Acero, 2/3 (January–March 1949), 4, footnote.
106 *Publicaciones*, Instituto del Hierro y del Acero, Madrid (January 1952), 18.
107 *Revista del Instituto del Hierro y del Acero*, CSIC-Instituto del Hierro y del Acero, 2/6 (October–December 1949).

108 'Homenaje al reverendo P. Vitoria', notice in *Revista del Instituto del Hierro y del Acero*, CSIC-Instituto del Hierro y del Acero, 8/35 (January–March 1955).
109 'S.E. el Jefe del Estado en nuestros laboratorios', *Revista del Instituto del Hierro y del Acero*, CSIC-Instituto del Hierro y del Acero, 8/36 (April–June), 121–4.
110 Insert in *Revista del Instituto del Hierro y del Acero*, CSIC-Instituto del Hierro y del Acero, 8/41 (October–December 1955).
111 The first edition of this encyclopaedia, popularly known in Spain as 'the Timan' [*sic*], had been published almost forty years earlier in 1910 in the United States by Hugh P. Tiemann, but its contents were still current; see Crespo García *et al.*, 2009, pp. 53–4.
112 The orders were to be sent to the institute headquarters at 15 Villanueva Street in Madrid; 'Boletín de pedido', *Revista del Instituto del Hierro y del Acero*, CSIC-Instituto del Hierro y del Acero, 4/3 (July–September 1951).
113 *CSIC*, 1949, 122.
114 Loose notice in the *Revista del Instituto del Hierro y del Acero*, CSIC-Instituto del Hierro y del Acero, 8/35 (January–March 1955).
115 *CSIC*, 1952, 141.
116 *CSIC*, 1954, 95.
117 *CSIC*, 1957, 97.
118 *CSIC*, 1959, 80.
119 *CSIC*, 1960, 91.
120 'La Revista por dentro', *INE: revista de información electrónica*, Madrid, Patronato Juan de la Cierva-Instituto Nacional de Electrónica, 7/26 (April 1964), 133.
121 *CSIC*, 1959, 85.
122 Even the first issue highlighted the 'urgent task of containing the escape of our experts towards what Cajal called "the anonymous heap of gold-diggers"' (p. 3). Two years later, in January 1960, Professor Barlo Vento confirmed this problem when he observed that 'what can clearly not be done is to let the human assets so laboriously moulded in the last twenty years be consumed by boredom and apathy. The exodus of Spanish researchers should make us think'; *INE: revista de información electrónica*, Madrid, Patronato Juan de la Cierva-Instituto Nacional de Electrónica, 3/9 (January 1960), 5.
123 *INE: revista de información electrónica*, Madrid, Patronato Juan de la Cierva-Instituto Nacional de Electrónica, 1/1 (January 1958), 3– 4.
124 For the advertisements published in 1958 and 1959, see *INE: revista de información electrónica*, Madrid, Patronato Juan de la Cierva-Instituto Nacional de Electrónica, 1–2/1–8 (January–October 1958–9).

125 *INE: revista de información electrónica,* Madrid, Patronato Juan de la Cierva-Instituto Nacional de Electrónica, 1/1 (January 1958), 35.

126 Arroyo Valdés was a colleague at the institute and head of Chemistry Laboratory at the INE's Department of Electronic Physics. She received the Juan de la Cierva Award for Team Research twice, in 1952 and 1959; see *INE: revista de información electrónica,* Madrid, Patronato Juan de la Cierva-Instituto Nacional de Electrónica, 6/21 (January 1963).

127 *INE: revista de información electrónica,* Madrid, Patronato Juan de la Cierva-Instituto Nacional de Electrónica, 1/2 (April 1958), 113.

128 *CSIC,* 1960, 96.

129 'La Revista por dentro', *INE: revista de información electrónica,* Madrid, Patronato Juan de la Cierva-Instituto Nacional de Electrónica, 7/26 (April 1964), 132.

130 'La Revista por dentro', *INE: revista de información electrónica,* Madrid, Patronato Juan de la Cierva-Instituto Nacional de Electrónica, 7/26 (April 1964), 132.

131 These regions contained the core of the recipients of a journal that, according to the CSIC – albeit without providing any evidence – also had 'subscribers in many other nations around the world'; *CSIC,* 1962, 107.

132 *CSIC,* 1963, 113.

133 *INE: revista de información electrónica,* Madrid, Patronato Juan de la Cierva-Instituto Nacional de Electrónica, 7/28 (October 1964), 352; *CSIC,* 1965, 128.

134 *INE: revista de información electrónica,* Madrid, Patronato Juan de la Cierva-Instituto Nacional de Electrónica, 7/28 (October 1964), 351.

135 *INE: revista de información electrónica,* Madrid, Patronato Juan de la Cierva-Instituto Nacional de Electrónica, 1/1 (January 1958), 31.

136 *CSIC,* 1962, 107.

137 *INE: revista de información electrónica,* Madrid, Patronato Juan de la Cierva-Instituto Nacional de Electrónica, 1/2 (April 1958), 99.

138 *INE: revista de información electrónica,* Madrid, Patronato Juan de la Cierva-Instituto Nacional de Electrónica, 4/13 (January 1961), 42.

139 *INE: revista de información electrónica,* Madrid, Patronato Juan de la Cierva-Instituto Nacional de Electrónica, 8/29 (January 1965), 56.

140 *CSIC,* 1965, 225.

141 *CSIC,* 1965, 225.

142 *CSIC,* 1951, 223.

143 'Editorial', *Revista de Plásticos,* Patronato Juan de la Cierva, 1/1 (January–February 1950), 1.

144 *CSIC, Memoria de la Secretaría General (1950)* (Madrid: CSIC, 1951), 31.

145 *Revista de Plásticos*, Patronato Juan de la Cierva, 1/1 (January–February 1950).
146 On the Gustavo Gili publishing house, see Fernando García Naharro, *Editando ciencia y técnica durante el franquismo. Una historia cultural de la editorial Gustavo Gili (1939–1966)* (Zaragoza: Prensas de la Universidad de Zaragoza, 2019).
147 Letter from José M.ª Aguirre (Gustavo Gili publishing house) to Joaquín de Oteyza, 12 July 1952; BC, Fons Editorial Gustavo Gili, Capsa 144, MS 9500/6, carpeta BARRON.
148 Letter from José M.ª Aguirre (Gustavo Gili publishing house) to Joaquín de Oteyza, 12 July 1952; BC, Fons Editorial Gustavo Gili, Capsa 144, MS 9500/6, carpeta BARRON.
149 Letter from Joaquín de Oteyza to Eduardo Fontseré, 15 July 1952; BC, Fons Editorial Gustavo Gili, Capsa 144, MS 9500/6, carpeta BARRON.
150 Letter from J. Fontán to Pedro Reverté, 17 July 1952; BC, Fons Editorial Gustavo Gili, Capsa 144, MS 9500/6, carpeta BARRON.
151 Letter from Salvador Terol Alonso to Pedro Reverté, 24 July 1952; BC, Fons Editorial Gustavo Gili, Capsa 144, MS 9500/6, carpeta BARRON.
152 Letter from J. Fontán to Pedro Reverté, 17 July 1952; BC, Fons Editorial Gustavo Gili, Capsa 144, MS 9500/6, carpeta BARRON.
153 Letter from the company Meryland S. R. L. to the Gustavo Gili Publishing House, 5 December 1952; BC, Fons Editorial Gustavo Gili, Capsa 148, MS 9500/6, carpeta COMPAÑÍA MERYLAND.
154 *Revista de Plásticos*, Patronato Juan de la Cierva, 1/1 (January–February 1950), 5.
155 For example, from the first issue: 'Cylinder kneading mill available, suitable for rubber or plastic materials, contact the Journal head office'; 'We buy all types of plexiglass, Perspex and polyvinyl chloride waste; write indicating price and send sample to the Journal head office'; *Revista de Plásticos*, Patronato Juan de la Cierva, 1/1 (January–February 1950), 11.
156 *Revista de Plásticos*, Patronato Juan de la Cierva, VII, June yearbook (1957), 1 and 144.
157 'Presentación', *Grasas y Aceites*, Instituto Especial de la Grasa y sus Derivados, *CSIC*, I/fasc. 1 (1950).
158 Editorial, 'Creación de la Asamblea de miembros del Instituto de la Grasa', *Grasas y Aceites*, Instituto Especial de la Grasa y sus Derivados, *CSIC*, VI/fasc. 6 (1955).
159 'Presentación', *Grasas y Aceites*, Instituto Especial de la Grasa y sus Derivados, *CSIC*, I/fasc. 1 (1950).

160 Editorial, 'Creación de la Asamblea de miembros del Instituto de la Grasa', *Grasas y Aceites*, Instituto Especial de la Grasa y sus Derivados, *CSIC*, VI / fasc. 6 (1955).
161 'Presentation', *Grasas y Aceites*, Instituto Especial de la Grasa y sus Derivados, *CSIC*, I / fasc. 1 (1950).
162 'Editorial', *Grasas y Aceites*, Instituto Especial de la Grasa y sus Derivados, *CSIC*, II / fascs 1–2 (1951).
163 *Grasas y Aceites*, Instituto Especial de la Grasa y sus Derivados, *CSIC*, IV / fasc. 2 (1953), 1.
164 José I Alonso Ramírez, 'Importancia de la información bibliográfica en la labor científica de nuestro Instituto', *Grasas y Aceites*, Instituto Especial de la Grasa y sus Derivados, *CSIC*, IV / fasc. 1 (1953), 43–8.
165 *Grasas y Aceites*, Instituto Especial de la Grasa y sus Derivados, *CSIC*, V / fasc. 4 (1954).
166 Editorial, 'Frente a la nueva campaña', *Grasas y Aceites*, Instituto Especial de la Grasa y sus Derivados, *CSIC*, VI / fasc. 5 (1955).
167 Editorial, 'Proyección internacional de nuestra revista', *Grasas y Aceites*, Instituto Especial de la Grasa y sus Derivados, *CSIC*, XI / fasc. 4 (1960).
168 Notice on the back pages, *Grasas y Aceites*, Instituto Especial de la Grasa y sus Derivados, *CSIC*, VI / fasc. 2 (1955).
169 *Grasas y Aceites*, Instituto Especial de la Grasa y sus Derivados, *CSIC*, IX / fascs. 1 and 2 (1958).
170 *Grasas y Aceites*, Instituto Especial de la Grasa y sus Derivados, *CSIC*, VII / fasc. 6 (1956).
171 *Grasas y Aceites*, Instituto Especial de la Grasa y sus Derivados, *CSIC*, IX / fasc. 1 (1958). In 1966, *Grasas y Aceites* was sent to the following countries: Germany (Democratic Republic), Germany (Federal Republic), Argentina, Austria, Belgium, Brazil, Bulgaria, Czechoslovakia, Chile, China, Colombia, Cuba, Egypt, the United States, France, Great Britain, Greece, Holland, Hungary, India, Israel, Italy, Japan, Morocco, Mexico, Paraguay, Peru, Poland, Portugal, Romania, Syria, Sweden, Switzerland, Tunisia, Turkey, the USSR, Venezuela and Yugoslavia; see Gómez Herrera, 1999, p. 193.
172 In 1957, *Chemical Abstracts* published 24,600 entries, 41.3 per cent more than the number published that same year in *Zentralblatt*; Michael Gordin, *Scientific Babel: How Science Was Done Before and After Global English* (Chicago IL: University of Chicago Press, 2015), p. 282.
173 Editorial, 'Proyección internacional de nuestra revista', *Grasas y Aceites*, Instituto Especial de la Grasa y sus Derivados, *CSIC*, XI / fasc. 4 (1960). '"*Grasas y aceites*" is one of the few journals specialising in fatty materials and in the early 1960s was given an ISO abbre-

viation for designation in the reference sections of research articles'; see Gómez Herrera, 1999, p. 192.

Chapter 12

1 F. S. Boig, 'Domestic and Foreign Periodicals in the Field of Organic Chemistry: A Statistical Analysis', *Science*, 110 (29 July 1949), 107.
2 *CSIC, Memoria de la Secretaría General (1950)* (Madrid: CSIC, 1951), 63.
3 Editorial, 'Unfortunately, the situation tends to perpetuate itself', *Grasas y Aceites*, Special Institute for Fats and their Derivatives, *CSIC*, XI, fasc. 1 (1960).
4 David M. Gates, 'Basic Research in Europe', *Science*, New Series, 128/3318 (1 August 1958), 233.
5 Editorial, 'Unfortunately, the situation tends to perpetuate itself'.
6 'It is shocking to see how people in positions of high responsibility respect and considerately listen to the opinions of scarcely-educated foreigners who represent nothing in their own country, as opposed to the contempt they display towards their national colleagues'; see Jaime Gracián Tous, 'Dificultades para el desenvolvimiento de la investigación en España', editorial in *Grasas y Aceites*, Special Institute for Fats and their Derivatives, *CSIC*, XI, fasc. 4 (1960).
7 'Letters', *Science*, New Series, 130/3368 (17 July 1959), 126–8.
8 Juan Comas Camps and Santiago Tarazaga Genovés, 'Science in Spain', *Science*, New Series, 130/3382 (23 October 1959), 1044. Juan Comas in particular would reiterate these criticisms starting in the 1940s; see Juan Comas Camps, 'Fobia Evolucionista', *Cuadernos Americanos*, 5/6 (1946), 66–73; and 'Reseña al *Manual de Antropología* de J. Pérez de Barradas', *Ciencia. Revista hispano-americana de Ciencias puras y aplicadas*, 9/7–10 (1948–9), 277–8.
9 *Blanco y Negro*, 13 September 1958, 1 November 1958, 6 December 1958 and 27 December 1958.
10 'Ciencia española', *ABC*, 25 November 1954.
11 The word 'objections ('*objeciones*') appears handwritten in blue ink over the word 'references' ('*alusiones*'), which is crossed out; AGUN/JIM/587/C-21 (1961).
12 Letter of 20 October 1961; AGUN/JIM/587/C-21 (1961).
13 AGUN/JIM/588/G-7 (1961).
14 'Franco y el Consejo Superior de Investigaciones Científicas, *ABC*, 1 October 1961.
15 This was, he said, 'slightly out of place, not because they might be outmoded, which is respectable, but because in some cases, they were legendary'; see Manuel Lora-Tamayo, 'La investigación

científica', an offprint from the second edition of *El Nuevo Estado Español* (Madrid: IEP-Editora Nacional, 1963), p. 13.

Chapter 13

1 *CSIC, Memoria de la Secretaría General (1945)* (Madrid: CSIC, 1946), 6. For the subsequent citations from the official Spanish National Research Council annual activity reports: *CSIC*, year of publication, and page number for the reference.

2 *CSIC*, 1942, 6.

3 *CSIC*, 1948, 56.

4 With Francisco Franco as its chief champion: 'It is logical – without logic lessening the credit and gratitude – that the man of initial and complete faith in the total triumph of Spain, the Caudillo of a rising Spain, is, directly and straight away, the chief champion of this national endeavour'; *CSIC*, 1942, 113.

5 *CSIC*, 1948, 151–2.

6 *CSIC*, 1942, 37.

7 *CSIC*, 1951, 77.

8 *CSIC*, 1952, 90–1.

9 The CSIC General Library and its specialist libraries took their place alongside the National Library, National Palace Library, El Escorial Library, Madrid National Conservatory Music Library, Madrid Ateneo Library, Menéndez Pelayo Library in Santander, Madrid Newspaper Library and Archive, National Newspaper Archive, Balaguer Library in Villanueva y Geltrú, Madrid Economic Society Library, Spanish-American Studies Library in Seville and the specialist libraries belonging to the royal and cathedral academies; see Amadeo Tortajada Ferrandis, 'Las bibliotecas en España. Su movimiento y estado actual', *Boletín de la Biblioteca General del CSIC. Boletín semestral*, 3–4, Year I (July–December 1949), 3–18.

10 *CSIC*, 1942, 281–5.

11 *CSIC*, 1944, 327–8 and 'La importante labor de las bibliotecas del CSIC', *ABC*, 14 January 1945.

12 Amadeo Tortajada Ferrandis, 'La Biblioteca General del Consejo', *Boletín de la Biblioteca General del CSIC. Boletín semestral*, 1–2, Year I (January–June 1949), 9–10.

13 'The Library Exchange Section now conducts exchanges with 998 Centres: 607 in Europe, 360 in the Americas and 31 in Africa and Asia. We receive thousands of journals from these exchanges, and this year [1945], 935 foreign non-periodical publications and texts have come in. Together with the acquisitions, this inflow has caused the Council libraries to grow significantly'; *CSIC*, 1946, 71–2.

14 *CSIC*, 1951, 86. To quote José Ibáñez Martín in April 1952, 'Books serve as a versatile and efficient instrument in all scientific research, and providing an abundant and convenient library collection has improved the work of the CSIC General Library. Thanks to an extremely valuable, daily contribution in recent years, it has reached its first 100,000 volumes, while the specialist libraries in the Institutes now hold more than one million volumes on their shelves'; *CSIC*, 1952, 83.

15 *CSIC*, 1943, 290.

16 'You can also request journals and books from any country, knowing that they will be handled just as quickly, with the expenses always borne by this General Library'; letter from Amadeo Tortajada to the director of the Institute of Physics and Chemistry, 14 May 1943; Rocasolano Library Archive, Carpeta Libros 1943.

17 *CSIC*, 1944, 313.

18 'General regulations for the free service of publications from the Spanish National Research Council to its member Institutes', 13 January 1943; Rocasolano Library Archive, Carpeta Libros 1943.

19 Order of 18 February 1941, *BOE*, 28 February 1941.

Chapter 14

1 This situation was similar to that of the university libraries in the country, which also needed to exchange publications with foreign institutions to compensate for the loss of their collections; letter from Javier Lasso de la Vega to José Cerezo Giménez, secretary of the Spanish Physics and Chemistry Society, 17 October 1939, BHU-CM AP 21/Caja 2/Carta 482b.

2 'They tell me that to complete the donation, they need our Ministry of Education to give express assurance that the subscriptions to these journals will continue from now on'; letter from the director of the Institute for Physics and Chemistry to the undersecretary of the Ministry of National Education, 1 May 1940; Rocasolano Library Archive, Carpeta Libros 1940.

3 Letter from José María Albareda to the director of the Alonso de Santa Cruz and Alonso Barba Institutes, 21 February 1941; Rocasolano Library Archive, Carpeta Libros 1941.

4 'Due to a lack of funds on our part, the library then requested that these be acquired through a request to the Embassy for the price of RM 3,035', note from the director of the National Institute of Physics and Chemistry to the undersecretary of National Education, 23 January 1940, Rocasolano Library Archive, Carpeta Libros 1939–40.

5 *ABC*, 16 April 1941, 12
6 Letter from the Rockefeller Foundation to Prof Julio Palacios, 18 April 1940; Rocasolano Library Archive, Carpeta Libros 1939–40.
7 Letter from the chief clerk of the centres and foundations, under the authority of the Undersecretary of the Ministry of National Education, to the director of the National Institute of Physics and Chemistry and the director's reply, 14 and 18 December 1939; Rocasolano Library Archive, Carpeta Libros 1939–40.
8 Of the first requests made by the National Institute of Physics and Chemistry for books and journals that rose to more than 18,000 pesetas, the undersecretariat of the Ministry of National Education granted them 6,000 pesetas in December 1939; letters of 2, 15 and 19 December 1939; Rocasolano Library Archive, Carpeta Libros 1939–40.
9 Letter from the American Chemical Society to the National Institute of Physics and Chemistry, 17 October 1940. Answered on 11 November 1940; Rocasolano Library Archive, Carpeta Libros 1939–40. See also publicity of 1 January 1941 in the Rocasolano Library Archive, Carpeta Libros 1941.
10 'Lista de revistas incompletas de la biblioteca de los institutos Alonso Barba y Alonso de Santa Cruz y gestiones hechas hasta la fecha para completarlas', 3 May 1941; Rocasolano Library Archive, Carpeta Libros 1941.
11 Letter from director Lyman J. Briggs, 21 June 1940; Rocasolano Library Archive, Carpeta Libros 1939–40.
12 3 May 1941; Rocasolano Library Archive, Carpeta Libros 1941.
13 Request letter, Rudolf Kadner, bookseller, 1 June 1943; Rocasolano Library Archive, Carpeta Libros 1943.
14 Letter from Amadeo Tortajada Ferrandis to the secretary of the Alonso Barba Institute, n.d.; Rocasolano Library Archive, Carpeta Libros 1943.
15 Rudolf Kadner, bookseller, Urgent!; Rocasolano Library Archive, Carpeta Libros 1943.
16 'Lista de revistas incompletas de la biblioteca de los institutos Alonso Barba y Alonso de Santa Cruz y gestiones hechas hasta la fecha para completarlas', 3 May 1941; Rocasolano Library Archive, Carpeta Libros 1941.
17 Letter from Amadeo Tortajada to the secretary of the Alonso Barba Institute, 27 October 1943; Rocasolano Library Archive, Carpeta Libros 1943. Two days later he wrote that various missing issues of *Ergebnisse der exakten Naturwissenschaften*, *ATM*, *Naturwissenschaften*, *Kolloid Zeitschrift*, *Physikalische Berichte*, *Zeitschrift für Physik* and *Zentralblatt für Mathematik* had arrived; 30 October 1943; Rocasolano Library Archive, Carpeta Libros 1941. A letter assumed to be written

by Julio Guzmán to Amadeo Tortajada in November 1943 said the following in this regard: 'I congratulate you and we congratulate ourselves on the formidable acquisition of English and American Journals just completed and whose receipt I acknowledge. With this, we are almost up to date, but so as not to function as beggars, I make note here of what we require of these journals to complete the collection, in the assurance that you will be able to fill the gaps for us'; 'Incomplete American and English journals in the institute library: *American Chemical Abstracts*, 1937, 31. Missing issues 19, 20, 21, 24 and years 1941, 42 and current; *British Chemical Abstracts*, the indices for 1935 and 1942; *Journal of the American Chemical Society*, 1937, Vol. 59, missing issue 11'; letter of 4 November 1943; Rocasolano Library Archive, Carpeta Libros 1943.

18 'Revistas inglesas recibidas por medio de la sección de intercambio', 30 June 1944; Rocasolano Library Archive, Carpeta Libros 1944.

19 *Biblioteca de los Institutos Alonso Barba y Alonso de Santa Cruz*, March 1945; Rocasolano Library Archive, Carpeta Libros 1944–5.

20 'Revistas inglesas recibidas por medio de la Sección de Intercambio', 30 June 1944; Rocasolano Library Archive, Carpeta Libros 1944–5.

21 Acquired and received on 14 October 1943. Letter from A. Tortajada to the director of the Institute of Physics and Chemistry; Rocasolano Library Archive, Carpeta Libros 1943.

22 'Libros ingleses que interesan a la biblioteca de los institutos Alonso Barba y Alonso de Santa Cruz', 2 June 1943; Rocasolano Library Archive, Carpeta Libros 1943.

Chapter 15

1 *Biblioteca de los Institutos Alonso Barba y Alonso de Santa Cruz*, March 1945; Rocasolano Library Archive, Carpeta Libros 1944–5.

2 Letter from Luis Cabrera Sánchez to the director of the Spanish National Research Council, Alonso de Santa Cruz Institute, 30 November 1945; Rocasolano Library Archive, Carpeta Libros 1945.

3 Antonio Madinaveitia Tabuyo, 'Don Blas Cabrera Felipe', *Ciencia. Revista hispano-americana de Ciencias puras y aplicadas*, 6/7–9 (1945), 241.

4 The following data come from three lists of books from Blas Cabrera's personal library that were either known to have been transferred to the CSIC or were missing from his collection: 'Libros de don Blas Cabrera depositados en la biblioteca del Instituto "Alonso de Santa Cruz" y que han sido entregados por este instituto a los herederos de dicho señor', 29 November 1945; 'Relación de los libros que faltan de la biblioteca particular de don Blas Cabrera', 30

November 1945 and 'Relación de libros de la biblioteca de don Blas Cabrera que han sido omitidos en la reclamación efectuada el 28 de noviembre 1945', 19 December 1945; Rocasolano Library Archive, Carpeta Libros 1945.

5 *Calcul des probabilités* (two volumes; n.d.), *La théorie de Maxwell et les oscillations hertziennes* (n.d.), *La Thermodynamique* (n.d.), *Leçons sur les hypothèses cosmogoniques* (n.d.), *Capillarité* (n.d.), *Les oscillations électriques* (n.d.), Électricité *et optique. La Lumière et les théories électrodynamiques* (n.d.) and *Polarisation et optique crystalline* (missing cover; n.d.); Rocasolano Library Archive, Carpeta Libros 1945.

6 *The Theory of Heat Radiation* (n.d.), *Eight Lectures on Theoretical Physics* (n.d.), *Vorträge über die kinetische Theorie der Materie und der Elektrizität* (n.d.), *Vorlesungen über die Theorie der Wärmestrahlung* (n.d.), *Einführung in die Theorie der Wärme* (n.d.), *Kasual Gesetz und Willensfreiheit* (n.d.), *Termodinámica* (n.d.), *Einführung in die Allgemeine Mechanik* (n.d.), *Einführung in die Theorie der Elektrizität und das Magnetismus* (n.d.) and *Physikalische Rundblicke* (n.d.); Rocasolano Library Archive, Carpeta Libros 1945.

7 *La structure de la lumière* (n.d.), *The Corpuscular Theory of Matter* (n.d.), *Electricity and Matter* (n.d.), *Rays of Positive Electricity* (n.d.), *Elements of the Mathematical Theory of Electricity and Magnetism* (n.d.), *Elettricittà e materia* (n.d.), *L´électron en Chimie* (n.d.) and *Conduction of Électricité through Gases* (n.d.); Rocasolano Library Archive, Carpeta Libros 1945.

8 *Ondes et mouvements* (n.d.), *Introduction a la physique des rayons X et gamma* (n.d.), *Sur une forme plus restrictive des relations d'incertitude* (n.d.), *Mécanique ondulatoire* (n.d.), *Théorie de la quantification dans nouvelles mécanique* (n.d.), *Une nouvelle conception de la lumière* (n.d.), *L'électron magnétique* (n.d.) and *Recueil d'exposés sur les ondes et corpuscules* (n.d.); Rocasolano Library Archive, Carpeta Libros 1945.

9 *Space, Time and Gravitation* (n.d.), *New Pathways in Science* (n.d.), *Espace, temps et gravitation* (n.d.), *Vues générales sur la théorie de la relativité* (n.d.), *Sur le problème de déterminisme* (n.d.), *Report on the Relativity Theory of Gravitation* (n.d.), *The Expanding Universe* (n.d.), *The Mathematical Theory of Relativity* (n.d.) and *The Internal Constitution of Stars* (n.d.); Rocasolano Library Archive, Carpeta Libros 1945.

10 *Cours de mécanique* (n.d.), *Cours de physique* (Part II; n.d.), *Cours de physique. Cours de thermodynamique* (n.d.), *Cours de physique. Cours de magnétisme et d´électricité. Applications* (n.d.), *Cours de physique. Électrooptique. Ondes Hertziennes* (n.d.), *Cours de physique. Étude des symétries* (n.d.) and *Mécanique. Essai des Matériaux* (n.d.); Rocasolano Library Archive, Carpeta Libros 1945.

11 *Théorie de la relativité* (n.d.), *Die Grundlagen der Allgemeine Relativitätstheorie* (n.d.), *The Meaning of Relativity* (n.d.), *Sur l'électrodynamique des corps en mouvement* (n.d.) and *Where is Science Going?* (n.d.); Rocasolano Library Archive, Carpeta Libros 1945.

12 *The Dynamical Theory of Gases* (n.d.), *Théorie dynamique des gaz* (n.d.), *The Mathematical Theory of Electricity and Magnetism* (n.d.), *Les nouvelles bases philosophiques de la science* (n.d.), *Astronomy and Cosmogony* (n.d.), *Report on Radiation and the Quantum-Theory* (n.d.) and *The New Background of Science* (n.d.); Rocasolano Library Archive, Carpeta Libros 1945.

13 *Problems of Atomic Dynamics* (n.d.), *Vorlesungen über Atommechanik. Vol. I* (n.d.), *La teoría de la relatividad de Einstein y sus fundamentos físicos* (n.d.), *Atomtheorie des festen Zustandes* (n.d.) and *La constitution de la matière* (n.d.); Rocasolano Library Archive, Carpeta Libros 1945.

14 Letter from Luis Cabrera Sánchez to the director of the Spanish National Research Council, Alonso de Santa Cruz Institute, 30 November 1945; Rocasolano Library Archive, Carpeta Libros 1945; see Fernando García Naharro, *Palabras, actores, objetos y lugares. El CSIC y la demarcación de la ciencia en la España franquista (1939–1966)* (Madrid: CSIC, 2022), pp. 335–45.

15 *Leçons sur la conductibilité des électrolytes*, by Eugène Darmois, was 'returned by hand to Mrs Cabrera', as was the second volume of *Curso de electrotecnia* by José Morillo y Farfán.

16 Letter from the director of the National Institute of Physics and Chemistry to the Commander of the General Staff of the Army First Corps, 20 June 1939; Rocasolano Library Archive, Carpeta Libros 1939–40.

17 Letter from the Commander of the General Staff of the Army First Corps to the director of the National Institute of Physics and Chemistry, 26 June 1939; Rocasolano Library Archive, Carpeta Libros 1939–40.

18 In April 1940, the institute officially communicated that it had received the private libraries of 'Messrs Moles and Madinaveitia' as a deposit; letter from the director of the National Institute of Physics and Chemistry to the head of the Library Recovery Service, 25 April 1940; Rocasolano Library Archive, Carpeta Libros 1939–40.

19 Letter from Vicente Navarro-Reverter to the director of the National Institute of Physics and Chemistry, 24 February 1940; Rocasolano Library Archive, Carpeta Libros 1939–40.

20 Javier Lasso de la Vega believed that José Giral's private library, which at that time was being held in the tourism offices at 4 Medinaceli Street in Madrid, should also be transferred to the Faculty of Pharmacy; letter from Javier Lasso de la Vega to Vicente Navarro-Reverter, 3 November 1939; BHUCM AP 21/ Caja 2/ Carta 802.

21 *Resultados de las experiencias hechas sobre el Alcanfor en Murcia*, Segovia, 1789; *Carta sobre los salitres*, Madrid, 1804 and *Análisis de una piedra meteórica caída en las inmediaciones de Lucena, en Aragón, el 14 de noviembre de 1773*, Madrid, 1804; 'Folletos prestados en esta Biblioteca (de Farmacia) al profesor E. Moles', 24 February 1940; Rocasolano Library Archive, Carpeta Libros 1939–40.

22 *Disertación sobre el Álkoli volátil, leída en las oposiciones públicas para las plazas vacantes de la Real Botica*, Barcelona, 1790? [*sic*]; *Plan de los ejercicios públicos de química que sostendrán los alumnos de la Escuela gratuita de esta ciencia, establecida en la ciudad de Barcelona por la Real Junta de Comercio del Principado de Cataluña, bajo la enseñanza y diseño del Dr. D. F. Carbonell y Bravo*, Barcelona, 1807; *Ejercicios públicos de química que sostendrán id. id.*, Barcelona, 1818; *Discurso que en la abertura de la Esc. gratuita de química, establecida en la ciudad de Barcelona por la Real Junta de Comercio del Principado de Cataluña, dijo el Dr. D. F. Carbonell y Bravo*, Barcelona, 1805; *Memoria químico-medica acerca de la preparación farmacéutica y usos medicinales del proto tartrato de mercurio y potasa. . . con lic*, Barcelona, 1832 and *Elogio histórico del Dr. Francisco Carbonell y Bravo, leído a la Academia de Ciencias Naturales y Artes de Barcelona en sesión extraordinaria de 3 de marzo de 1838*, Barcelona, 1838; 'Folletos prestados en esta Biblioteca (de Farmacia) al profesor E. Moles', 24 February 1940; Rocasolano Library Archive, Carpeta Libros 1939–40.

23 Letter from the director del National Institute of Physics and Chemistry to the director of the Library Return and Recovery Services, 1 March 1940; 'Folletos prestados en esta Biblioteca (de Farmacia) al profesor E. Moles', 24 February 1940; Rocasolano Library Archive, Carpeta Libros 1939–40.

24 Specifically, 645 books, according to the inventory conducted in 1939 by the interim secretary of the National Institute of Physics and Chemistry, Mr Garrido, on the request of the interim director, Prof Julio Palacios; 'Libros pertenecientes a la biblioteca del Dr. E. Moles'; Rocasolano Library Archive, Carpeta Libros 1948; see Fernando García Naharro, *Palabras, actores*, pp. 347–76.

25 'Lista de libros que se devuelven al Dr. E. Moles', 19 November 1948; 'Libros pertenecientes a la biblioteca del Dr. E. Moles'; Rocasolano Library Archive, Carpeta Libros 1948.

26 'We also have a room set aside for storing offprints of the works done in these Institutes and published in the journal "Anales de Física y Quimia"'; *Biblioteca de los Institutos Alonso Barba y Alonso de Santa Cruz*, March 1945; Rocasolano Library Archive, Carpeta Libros 1944–5.

27 'Lista de libros que se devuelven al Dr. E. Moles, 19 de noviembre de 1948', Rocasolano Library Archive, Carpeta Libros 1948. In a let-

ter sent to José María Albareda in December 1950, Enrique Moles continued to ask the CSIC 'to return some 300 volumes, the exact list of which was drawn up in 1939'; see Sales, *Enrique Moles: una biografía científica y política* (Barcelona-Madrid: Universitat de Barcelona-CSIC, 2021), p. 339.

28 'The students who pass the *Verbands* test with him can boast of their knowledge of chemical analysis'; see Enrique Moles Ormella, 'Un curso teórico y práctico de Química-física', *Anales de la JAE*, IV, Memoria 2 (1911), pp. 67–90.

29 'Una conversación con el químico señor Moles', *El Sol*, 20 June 1927; cited in Thomas F. Glick, *Einstein y los españoles: ciencia y sociedad en la España de entreguerras* (Madrid: CSIC, 2005), p. 38.

30 Other scientists from the geographic periphery whose books were in his library included the eminent Bengali chemist Parfulla Chandra Ray (1861–1944), the Western-educated Persian Abdul-Chalig Achundow and the Romanian chemist Gheorghe Gh. Longinescu (1869–1939).

Chapter 16

1 'One *Zeitschrift f. Kristallographie* was lost en route and another due to the effect of the bombing'; note by Rudolf Kadner, September 1944; Rocasolano Library Archive, Carpeta Libros 1944–5.

2 Rocasolano Library Archive, Carpeta Libros 1946.

3 The library was open from 10 a.m. to 1 p.m. and 4 p.m. to 6 p.m., Monday to Saturday; see *Biblioteca de los Alonso Barba and Alonso de Santa Cruz Institutes*, March 1945; Rocasolano Library Archive, Carpeta Libros 1944–5.

4 Letter from the managing director of La Madrileña S.A. to the director of the National Institute of Physics and Chemistry, 4 September 1944; Rocasolano Library Archive, Carpeta Libros 1944–5.

5 Letter from Fernando Martín to Julio Palacios, 29 June 1947; Rocasolano Library Archive, Carpeta Libros 1947.

6 In any case, Julio Palacios answered the writer, specifying that the reference to the article must be wrong, since it was not in the cited journal, and indicating that if he would correct the mistake and make the request again, he would have 'a photocopy service via which you can obtain the article that interests you' at his disposal; letter from Julio Palacios to Fernando Martín, 10 July 1947; Rocasolano Library Archive, Carpeta Libros 1947.

7 Letter from Manuel Colomina to Raimundo Menéndez, 16 April 1948; Rocasolano Library Archive, Carpeta Libros 1947.

8 Circular signed by Amadeo Tortajada, sent to the director of the Alonso Barba Institute, 16 June 1944; Rocasolano Library Archive, Carpeta Libros 1944.

9 Letter to the director of the Alonso Barba and Alonso de Santa Cruz Institutes, 2 October 1945; Rocasolano Library Archive, Carpeta Libros 1945.

10 *Biblioteca de los Alonso Barba and Alonso de Santa Cruz Institutes,* March 1945; Rocasolano Library Archive, Carpeta Libros 1944–5.

11 This version of the work had been translated into Spanish in 1932 and included a long chapter on wave and quantum mechanics; see Marià Baig i Aleu, Gonzalo Gimeno Valentin-Gamazo and Mercé Xipell Gómex del Moral, 'La introducción de la mecánica cuántica en España: las primeras lecciones y los primeros textos', in Nestor Herran and Xavier Roqué (eds), *La física en la dictadura. Físicos, cultura y poder en España 1939–1975* (Bellaterra: UAB, 2012), p. 164.

12 'Relación de los libros que faltan de la biblioteca particular de Don Blas Cabrera'; Rocasolano Library Archive, Carpeta Libros 1944–5.

13 At the time, almost all the works by Spanish authors or translated into Spanish were also missing, but their whereabouts were unknown. 'Revisión de las obras que faltan de la Biblioteca de Enrique Moles', March–June 1946; Rocasolano Library Archive, Carpeta Libros 1948.

14 Letter from Julio Palacios to Charles Roos, 29 November 1945, Rocasolano Library Archive, Carpeta Libros 1944. Regarding the purchase of eight books directly from W&G Foyle Ltd (London), see the letter dated 9 July 1947, Rocasolano Library Archive, Carpeta Libros 1947.

15 In addition to *Society Automotive Engineering Journal*, 39 (1936). 'Libros pedidos por D. Julio Palacios'; Rocasolano Library Archive, Carpeta Libros 1946 and 1947.

16 Letter written from the Alonso de Santa Cruz Institute to the director of the CSIC Publications Section, unsigned, 12 December 1945; Rocasolano Library Archive, Carpeta Libros 1944.

17 Letter from Amadeo Tortajada to the deputy director of the Alonso Barba Institute, 'Mr Lora-Tamayo', 5 September 1946; Rocasolano Library Archive, Carpeta Libros 1946.

18 In addition: 'Only requests for certain special works for the independent service of each of these centres can go directly through the directors of the Institutes, and they must be financed by the Institute budget'; circular from Amadeo Tortajada, 30 October 1947; Rocasolano Library Archive, Carpeta Libros 1947.

19 Letter from Manuel Colomina to the publishing house Heffer & Sons Ltd (Cambridge, England), 19 November 1948; Rocasolano Library Archive, Carpeta Libros 1947. Although the collaborator Miguel Belenguer received permission to take care of some books and journals

for the library while he was travelling to London in 1947, he clearly continued to receive other orders in his name for some time.

20 Rocasolano Library Archive, Carpeta Libros 1949–53.

21 Rocasolano Library Archive, Carpeta Libros 1953–60

22 Amadeo Tortajada Ferrandis, 'Las bibliotecas especiales. Estudio particular de las del Consejo', *Biblioteca General del CSIC. Boletín semestral*, 2/5 (January–June 1950), 16.

23 Javier Lasso de la Vega, 'Los progresos de la producción científica de lengua española y el acceso a la literatura técnica impresa en caracteres cirílicos', *Biblioteca General del CSIC. Boletín semestral*, 5/12 (July–December 1953), 157.

24 Nonetheless, the report also noted that the 'Physics and Chemistry library in the "Alonso Barba" and "Alonso de Santa Cruz" Institutes can be considered the best in Spain within its specialty'; *CSIC, Memoria de la Secretaría General (1952–1954)*, II (Madrid: CSIC, 1958), 1214.

25 Letter from José Ibáñez Martín to Mariano Navarro Rubio, 22 December 1961; AGUN/JIM/139/587/C-21 (1961).

26 Letter from José Ibáñez Martín to Mariano Navarro Rubio, 22 December 1961; AGUN/JIM/139/587/C-21 (1961).

Conclusion

1 *BOE*, 16 December 1966. See Manuel Lora-Tamayo, 'Restructuración electiva del Consejo', in Manuel Lora-Tamayo, *Lo que yo he conocido. Recuerdos de un viejo catedrático que fue ministro* (Cádiz: Federico Joly y Cia., 1993), pp. 270–3. All these changes were promoted by Manuel Lora-Tamayo, as Minister of Education, who also created the fund for the Promotion of University Research with a committee of university professors responsible for managing its distribution, as well as the National Fund for the Development of Scientific Research, which had an initial budget of 100 million pesetas. Decades later, historians José Luis Peset and Elena Hernández-Sandoica would write that the inception of the entire Spanish research system, which formally culminated in the 1986 Science Law, can be found in the development of Lora-Tamayo's innovations; see José Luis Peset Reig and Elena Hernández Sandoica, 'La recepción de la cultura científica en la España del siglo XX: la Universidad', in Antonio Morales Moya (coord.), *Las claves de la España del siglo XX. Vol. 8. La cultura* (Madrid: España Nuevo Milenio, 2001), p. 145.

2 Letter from Rudolf Kadner to the National Institute of Physics and Chemistry, 31 August 1940; Rocasolano Library Archive, Carpeta Libros 1940.

3 Letter from Alfredo Sánchez Bella to the secretary of the Alonso Barba and Alonso de Santa Cruz Institutes, 19 December 1940; Rocasolano Library Archive. Carpeta Libros 1939–40.

Epilogue

1 Presentation by Eloísa del Pino (CSIC president), De Organización y Relaciones Institucionales, CSIC-Vicepresidencia. *El tratamiento de la memoria democrática en el CSIC. Informe sobre la adaptación a la ley 20/2022, de 19 de octubre, de memoria democrática* (Madrid: CSIC – Vicepresidencia de Organización y Relaciones Institucionales (VORI), 2024), 7 *https://doi.org/10.20350/digitalCSIC/16160*. For the subsequent citations from the official CSIC democratic memory report: *El tratamiento de la memoria,* and page number for the reference.

2 The names of other researchers who have appeared on the pages of this book, such as Manuel Lora-Tamayo, Eduardo Torroja, Alberto Sols and Antonio Rius Miró, were also examined, with the decision made in the end to maintain their names; although their research careers coincided with the Franco period, their actions did not violate the stipulations in the Law of Democratic Memory. *El tratamiento de la memoria,* 21–5.

3 A few years earlier, in 2020, the library of the CSIC Institute for Physical Chemistry changed its name to the Dorotea Barnés Physics and Chemistry library in honour of the doctor in chemistry who worked with Miguel Catalán in the spectroscopy section of the National Institute of Physics and Chemistry. The new library resulted from the unification of the libraries of the Rocasolano Institute for Physical Chemistry (IQFR) and the Miguel Catalán Centre for Physics, which included the Institutes for Material Structure (*Institutos de Estructura de la Materia,* IEM), the Daza de Valdés Optics Institute (*Instituto de Óptica Daza de Valdés,* IO) and the Institute for Fundamental Physics (*Instituto de Física Fundamental,* IFF).

4 *El tratamiento de la memoria,* 25.

5 *El tratamiento de la memoria,* 29.

6 *El tratamiento de la memoria,* 30.

7 On 13 July 1967, the CSIC approved the appointment of Ibáñez Martín as honorary president in perpetuity ('*con carácter perpetuo*'). However, the 2024 CSIC report argues that 'the Royal Spanish Academy equates the adjective "in perpetuity" (*perpetuo*) with "for life" (*vitalicio*) when applied to positions, and defines this adjective as "until the end of life". Given that Ibáñez Martín died on 31 December 1969, the appointment is understood to have expired and no formal revocation is necessary'. In this vein, the CSIC also made a commitment to review and revoke all the medals and decorations

that the council had given to those who had 'formed part of the repressive apparatus of the Franco dictatorship'. *El tratamiento de la memoria*, 47; 60.

8 *El tratamiento de la memoria*, 29.

9 *El tratamiento de la memoria*, 45.

10 The CSIC commemorates eighty-five years of research at the service of society, *www.csic.es/es/actualidad-del-csic/el-csic-conmemora-85-anos-de-investigacion-al-servicio-de-la-sociedad* (last accessed 6 January 2026). Act: 85th Anniversary of the CSIC and presentation of the CSIC Extraordinary Medal for Scientific Merit, *www.youtube.com/watch?v=ONuZV75XUWY* (last accessed 6 January 2026).

11 *El tratamiento de la memoria*, 41.

12 *El tratamiento de la memoria*, 41.

Index